SCIENCE AND POLITICS

JEAN-JACQUES SALOMON

SCIENCE AND POLITICS

Translated by Noël Lindsay

THE M.I.T. PRESS
Cambridge, Massachusetts

First published 1973 by
THE MACMILLAN PRESS LTD
London and Basingstoke
Associated companies in New York
Dublin Melbourne Johannesburg and Madras

First M.I.T. Press edition September 1973

Library of Congress catalog card number 73–2054
ISBN 0 262 19111 3

Printed in Great Britain

for
CLAIRE

Contents

Contents

Introduction

I

This book is about politics—about political decision and action on issues which affect science. Politics as a science in its own right is considered only as a secondary topic, though it is of necessity considered; the two lines of thought inevitably converge, since matters of science, by becoming a political issue, in turn affect the exercise of power. If the subject is defined as 'science in its relations with politics', it is still ambiguous. The ambiguity would have been even greater if the title chosen had been 'science policy' or, as it is often called, as inaptly in English as in French, 'scientific policy'. Strictly speaking there is no single science policy and a policy is no more 'scientific' than any other policy merely because it deals with issues affecting science; there are policies whose subject matter is scientific and technical research, its development and the exploitation of its results, and these policies are merely one aspect among others of a country's general policy.

The word 'politics' itself is so vague and general that we are duty bound to define the exact meaning we give it, however arbitrarily.[1] In the literature on this subject, the English language has the advantage over French that it has two words, 'policy' and 'politics', to denote two sharply distinguished aspects of what a French writer is compelled to call *la politique*; there is a 'science policy' which is the whole body of thinking, action and action programmes about science, and there is a 'politics of science' which is the field in which science comes into contact with politics—and with the politician. But this advantage is only apparent, since politics regarded as the technical theatre of decisions and

1. 'The word "politique" in French has a variety of meanings. We speak of domestic policy, of foreign policy, of Richelieu's policy and of the policy on alcohol or beets; and sometimes it seems impossible to find the connecting link between these different usages. Bertrand de Jouvenel in a recent book has said that the word is used to convey such different meanings that we had simply better decide for ourselves the meaning which we want to give it.' Raymond Aron, *Democracy and Totalitarianism* (trans. Valence Ionescu), Weidenfeld and Nicolson, London (1970), 3.

actions cannot be divorced from politics regarded as the philosophical theatre of conflicts and rivalries—the politics of science no more than any other branch of politics, and precisely *because* it deals with science.

We therefore define politics in its broadest sense, as Raymond Aron had in mind when he wrote

> 'Politics' is a translation of the Greek word *politeia*. Politics is essentially what the Greeks called the way in which the city is run, that is, the method of establishing command, taken as characteristic of the method by which the entire community is run.[2]

We are all the more prepared to accept the ambiguity of the term 'politics of science' since there is now a two-way relationship between science and politics, between knowledge and power, between statesmen and researchers. Science is no less bound up with the polity than the polity is bound up with science.

The definition of science, on the other hand, would be less ambiguous if we accepted the common view of science as a body of knowledge and results which, because they are founded on methods of experiment and verification, are in theory universally recognised. But things are not as simple as that. In the first place, this knowledge and these results cannot be divorced from the activities which nourish them, on which they depend and which at the same time they make possible; furthermore, the term 'science' designates an object of which this knowledge and these results are still only the signs or, if you like, in archaeological language, the monuments; neither a philosophical category nor a scientific concept, but an ideological notion. As Louis Althusser has said, there is no 'science', there are only 'sciences'.[3] We leave aside the epistemological question of what these sciences are, and concentrate instead on the way in which they form a social institution, the practice of which is known as scientific and technical research. In this sense, science is the activity carried on by researchers—scientists, engineers and technicians—in the context of knowledge, methods, procedures and techniques sanctioned by experiment and verification.

This activity takes so many forms because it is carried on for such varied motives and in such varied organisations, and above all because it is concerned with such different disciplines, that the formula 'scientific and technical research' seems convenient rather than adequate to describe the unity among the various functions it embraces. This is even

2. *Ibid.*, 6.

3. L. Althusser, thesis 26 of Philosophy Course for Scientists, given at the Ecole Normale Supérieure, Paris, 1967–8 (stencilled).

more true if a distinction is drawn within the activities between their different levels, ranging from the widest theoretical generalisation to the narrowest application. The whole practice of research tends to disregard these distinctions, which are grounded on an outdated conception of science rather than on any intellectual hierarchy among the different forms of research. The specific characteristic of modern scientific research is not only that it is associated with technology, of which it is the prerequisite and the determinant, but also that it increasingly abolishes any lack of continuity between the generalisation stages and the application stages of the process of discovery and invention. It is for this good reason that, recognising that no sharp distinction can be drawn between fundamental research and applied research, we talk in this book of *the scientific research system* as it now appears to us: defined by the objectives of the institutions in which the researchers carry on their activities, rather than by the nature of these activities or by the motives, attitudes and objectives of the researchers themselves.

To avoid any ambiguity, however, science is above all conceived here in the narrow sense which the word bears in the English language, of the exact and natural sciences as distinguished from the social and human sciences. This is not to say that, in common practice as well as in theory, politics are not affected by the social and human sciences and *vice versa*. In some countries these disciplines are placed on an equal footing in the name of the unity of knowledge, the concept of science embodying, as in Germany (*Wissenschaft*) or Russia (*nauka*), all disciplines which are the subject of university teaching or research; in a great many countries the support accorded to the social and human sciences passes through the same government channels as support for the exact and natural sciences, and wherever the sciences of man are not yet represented in the science policy system, there is a manifest tendency towards integrating them into it. This book, however, is not directly concerned with state use of the social and human sciences and their influence on public affairs, even if there is every reason to regard such developments as another aspect of the same evolution by which knowledge, summoned to give proof of its utility and financed *pro rata* to its applications, enlists itself in the service of power.[4] We shall come back to this point, but in the meantime when we speak of science we mean,

4. See 'Problèmes posés par une étude des sciences sociales et humaines', *Revue internationale des Sciences Sociales* (UNESCO, Paris), 16, no. 4 (1964), and especially the contribution by Claude Levi-Strauss, 579–97; *Social Sciences and the Policy of Governments*, OECD, Paris (1966), and the series

unless otherwise specified, the exact and natural sciences, without entering the debate on the unity of scientific knowledge or the more or less scientific character of one field compared to another, and we shall refer to the special problems of the social and human sciences only so far as they help to throw light on the problems raised by the relation of the exact and natural sciences with policy.

When political authorities treat science as a means, we may ask whether they are not diverting it from its true end. But knowledge itself treats power as the instrument of its progress and we are driven to ask whether the autonomy of which it boasts is not an illusion relating back to a state of science which no longer exists, or, going still deeper, whether the end which it pursues does not spring from a purpose which has no content. In this dialectical relation between science and the authorities, it is idle to question which of the two makes use of the other, since the services rendered are quite manifestly reciprocal, but the question how far their objectives are identical takes us to the heart of the problem: is science destined to become the servant of the state because state support depends on the services the state expects from science, and because science can see no other way of making progress except by winning more and more favours from the authorities? So long as science, even defined solely as an instrument, had not given proof of its power—so long as the passage from pure research to application was regarded merely as a possibility—the question of the political use of scientific research could have been shelved as being academic, or framed in terms of mythology, from the Golem to Frankenstein by way of Goethe's Faust; the sorcerer's apprentice was still a sorcerer rather than professionally invested with the means of affecting political power. Scientific research aspired to win the support of the state while holding its institutions at arm's length; it could well propose itself, in harmony with the philosophy of the Enlightenment, as the best means towards the very ends which were assigned to power—the progress, well-being and happiness of mankind—but it merely tendered its services on the condition of keeping control over its own affairs.

On the morrow of the second world war, the future of science was bound up with the future of the political authorities; the question of the political use of scientific research arose in the light both of the growing needs of research and of the influence which research results exerted over

---

of reports published by the National Academy of Sciences on research in the social sciences, particularly *The Behavioral and Social Sciences, Outlook and Needs*, Washington (1969).

the affairs of politics itself. Science can no longer be divorced from the political programme which allows it to develop, and which science itself helps to achieve through close association with the decisions that determine the programme and fix the ways in which it can be achieved. Even though the practice of 'pure' science seems unchanged, in the sense that the initiative and methods of a research project in no way depend on instructions from any other source than the curiosity or interest of the men who decide to devote themselves to it, the choices affecting that project are set in a context of institutions, mechanisms, procedures and decisions of which it is no longer the sole master. The government of science may be more or less liberal, flexible and decentralised, but in the last analysis it always depends on the political decision-making system which foots the bill.

The problems raised by science policy lie at the interface of a number of disciplines—economics, sociology, political science, history of science, not to mention the contributions which scientists themselves never stop making. It is not surprising that a common approach to these problems should be advocated; at university level such an approach is desirable, and even essential. Some universities—in Britain, Sweden, Russia and above all in the United States—have already set up interdisciplinary units of this kind, and some people go so far as to speak of a new discipline, 'the science of science', which would group together all disciplines that could help towards a better understanding of the problems inside and outside the contemporary enterprise of scientific and technical research. This field is new, certainly, because little is known about it; on such questions as the economics of scientific research, the psychology of researchers, the weight of science in foreign policy, or the role of scientists as government advisers, the surface is barely scratched. The need for specialised studies proves how far we are from any kind of synthesis, but the multiplication of such studies also shows that this is the direction to follow. The artificer in the field of reason, says Kant, must be followed by the lawgiver of reason: a person who cannot be found in practice, but whose function is one that must be performed since it deals with 'the relation of all knowledge to the essential ends of human reason'.[5]

In this sense the approach adopted in this essay is primarily philosophical, and an account of its origin may perhaps help the reader to put it in its proper setting. Its prime source is an event; the perfection of the first atom bombs and the 'Oppenheimer case', that is to say, the

5. I. Kant, *Critique of Pure Reason* (trans. Norman Kemp Smith), Macmillan, London (1958): 'Architectonic of Pure Reason', 657–8.

realisation of the new role of the scientist in the modern world and the
pitfalls involved in his collusion with power. The second source is a
conceptual analysis, a lecture by our teacher Georges Canguilhem on a
theme which, more than ten years ago, went to the heart of the difficulties we are at present experiencing: does not the whole success of
modern science tend to warrant nihilism, just as the challenge to the
legitimacy of science, at a time when it had little or no mastery over
nature, opened the way to scepticism? Science could be doubted as a
method when it did not aim at action; in our day, while the results
achieved by this method are not open to any doubt, they are not thereby
endowed with reason. Science has proved itself as a means; it no longer
tries to prove itself as a philosophy. As it has developed over the last
quarter of a century under state patronage, the scientific enterprise has
become, with the university, the most revealing scene of 'the crisis of
civilisation'. It is no accident that the challenge facing science today
coincides with the indictment of industrial societies which, dizzy from
their technical conquests, are once again discovering the human and
social cost of progress.

It was undoubtedly Husserl who first and most dramatically became
aware of this; there is a close relationship between science, reduced to
the evidence of its results, and 'the bankruptcy of humanism'. The more
efficient science is, the less it answers the question of the meaning or lack
of meaning of human existence; the more 'rewarding' it is, the less it
seems to serve humanism. The exploits of the most recent technology—
which, as will be seen, seem to us no longer distinguishable from the
conquests of science—merely make more manifest the emptiness of the
faith they encourage in the power of reason to solve the problems of
mankind.

Husserl's indictment applies no less to the atomic or space age than
to its predecessor:

> We make our beginning with a change which set in at the turn of the past
> century in the general evaluation of the sciences. It concerns not the
> scientific character of the sciences but rather what they, or what science in
> general, had meant or could mean for human existence. The exclusiveness
> with which the total world-view of modern man, in the second half of the
> nineteenth century, let itself be determined by the positive sciences and be
> blinded by the 'prosperity' they produced, meant an indifferent turning
> away from the questions which are decisive for a genuine humanity.
> Merely fact-minded sciences make merely fact-minded people.[6]

6. E. Husserl, *The Crisis of European Sciences and Transcendental Phenomenology* (trans. David Carr), Northwestern University Press, Evanston,
Illinois (1970), 5–6.

II

Criticism of modern societies oscillates between the vain image of a return to primitive equilibrium and the messianic vision of the transition to some completely new equilibrium. Here is proof, at any rate, that disequilibrium exists, but also proof that rejecting the order whose organisation and tendencies shock us is not in itself enough to bring back the old order or to establish a new one; the choices which pass as the most reasonable are not made solely because they are more rational. The image of a society capable of transcending what in our society is unjust, absurd or fatal still has to be worked out, but on condition that we first recognise that there is no escape from the conquests of science and technology. Utopia does not change the world if it postulates an impossible return to the past; it can, however, help to define the vision of a different future if it takes account of the servitudes of the present. Science and technology are our destiny and we can do nothing except start by learning to guide the direction of their future.

This essay does not profess to give the recipe for such an apprenticeship; starting from a limited analysis of one facet of contemporary reality, it tries to understand how we got to this point and the direction which can be given to the irreversible collusion between science and power. Neither totally critical nor totally acquiescent, the analysis certainly aims at realism, since it does no more than look at the course of events to find out just what made them inevitable; but it also aims at being a philosophical discourse, since it does not abandon hope of finding a meaning, while recognising that it does not hold the key either of truth or of salvation. In sum, this essay is an investigation of the irrationality of the institution which most strikingly embodies the rationality of the West.

If it has a theme, it is that the relations established between science and power in the mid-twentieth century were inherent from the outset in the whole nature of modern science; there is no need to indict the economic form of modern societies, their political régimes or their ideologies to account for the alienation of the researcher from the state. The essence of contemporary science is that it is both organised research and the deliberate exploitation of its results; in Heidegger's words, 'the apparition of the gigantic' as a sign of the process whereby science has become 'the conquest of the world as an image conceived' [7]—gigantic

7. M. Heidegger, 'L'époque des conceptions du monde', in *Chemins qui ne mènent nulle part*, Gallimard, Paris (1962), 85.

in the direction of the largest, the most remote, the smallest—has been made possible only by the attention paid to science *fulfilled as technology*. The gigantic is the sign not only of the future of science, but also of the will of society.

Far from being fortuitous, this convergence seems to us to result from the very nature of modern science; by conceiving nature from the horizon of instrumentality, science condemned itself to be no more than an instrument. In its relations with power, 'pure' research professes to be something which it is not, while it is merely fulfilling a destiny which cannot be divorced from the destiny of society: *just as the frontier between pure science and applied research is blurred, the frontier between knowledge and its application no longer exists.* It is convenient to speak of the scientist as a 'pure' researcher, alienated by his subservience to the power on which his continued work depends. But is there still any such thing as 'pure' research? In this sense, the alienation of the scientist does not seem to us to arise from the type of society which encourages his research in greater or lesser degree or with more or less freedom, but to arise from a conception of science whose ends are in no way alien to the social system which subsidises its progress.

By the very fact of making knowledge a source of power, modern science condemned itself to ambivalence; knowledge of the universe and mastery of nature constitute two functions which the scientific revolution of the seventeenth century had precisely defined as inseparable from each other. But the time lag between disinterested knowledge of the universe and its practical applications, and the difficulty of passing from one to the other on a large scale, for a long time prevented any question being raised about the use of pure knowledge as a source of power over nature and men and its consequences. What had merely been postulated in the optimistic perspective of the mastery of the powers of science by science itself, came into being in the dramatic context of a political use of pure knowledge which science was to claim was against its own will.

> The problems scientists are called on to solve are for the most part selected by the scientists themselves. For example, our Defense Department did not one day decide that it wanted an atomic bomb and then order the scientists to make one. On the contrary, it was Albert Einstein, a scientist, who told Franklin D. Roosevelt, a decision-maker, that such a bomb was possible.[8]

From the outset, therefore, science takes shape as a political act and a political problem; the political, strategic or diplomatic aspects of certain

8. A. Rapoport, 'The Use and Misuse of Games Theory', in *Science, Conflict and Society*: readings from *Scientific American* (1969), 286.

scientific research can no longer be divorced from its technical aspects. In the decision-making process there are no longer any distinct frontiers between the sphere of the politician and the sphere of the scientists; in some instances the frontier is so fine that the power of decision in fact lies with the scientists on political questions and with the politicians on scientific questions. Science tenders its services to power and becomes a partner in its decisions; power makes use of science and becomes a partner in its destiny.

But this turning point in the relations between science and politics does not mean that science has turned its back on one vocation in order to alienate itself in favour of a new cause; it is in fact a realisation of the potentialities which have been present in modern science from its origins. The break is no revolution, but the culmination of a movement foreshadowed by the seventeenth century, dreamed of by the eighteenth century and adumbrated in the nineteenth. 'The break of day that, like lightning, all at once reveals the edifices of the new world,' said Hegel, introduces a qualitative leap into 'the continuity of purely quantitative growth'.[9] Just as the ambivalence of modern science was inherent in its nature, the relations between science and politics were condemned to ambiguity by the very fact that the progress of science was to depend more and more on state support. Atomic research did not add fresh ambiguity to a situation which the researchers themselves had always wished for; but it was the *product* of the conjunction which took place, on the occasion of the second world war, between the interests of science and the interests of the state, with all its consequences for the exercise of science and of power, which was to multiply the new factors in this situation.

While many scientists would prefer, as Don K. Price recalls, not to have to live off government largesse, they have no alternative; they are rather like the young lady of Kent

> Who said that she knew what it meant
> When men asked her to dine,
> Gave her cocktails and wine;
> She knew what it meant—but she went.[10]

This situation evokes, in more philosophical terms, Sartre's female hypocrite who does what she professes to eschew and gives what, from

9. Hegel, 'Preface to the Phenomenology', in W. Kaufmann, *Reinterpretation, Texts and Commentary*, Doubleday, New York (1965), 380.
10. Don K. Price, *Government and Science* (1953), reissued by Oxford University Press (1962), 87.

her public behaviour, she might be expected to refuse.[11] And so we find scientists expressing nostalgia for the 'good old days' when science was not condemned to ambiguity, just as though it had not always been inclined from the start to fulfil itself with state aid:

> But after the idyllic years of world science, we passed into a tempest of history; and by an unfortunate coincidence, we passed into a technological tempest, too.[12]

Even more significantly, this nostalgia is found in the Soviet Union no less than under the liberal régimes, thus confirming how little, before the war, the control exercised by the Communist state subordinated pure science to political pressure:

> The year that Rutherford died there disappeared for ever the happy days of free scientific work which gave us such delight in our youth. Science has lost her freedom. Science has become a productive force. She has become rich but she has become enslaved and part of her is veiled in secrecy.[13]

By the same token, the dependence of scientific research on the public authorities brings the political régimes closer together in spite of all that separates them: the 'advanced' industrial society is the one which involves the intervention of the state in the affairs of science, nationalising or socialising, little by little, a great part of the economy, in spite of the dogma that in a capitalist economy the market shall be ruled by competition only; research activities have taught the United States Federal Government 'how to socialise without assuming ownership'.[14] Just as the reputedly sovereign consumer has no initiative in deciding what should be produced, so the research system does not depend solely on the initiative of the reputedly autonomous scientist. It is neither abundance nor the aspiration to abundance which brings the economic systems of industrialised societies into line with one another, but the requirements of science and technology as formulated through the filter of state imperatives—national defence, support of growth industries, prestige. The ambiguity of relations between science and politics is manifested more than ever in this convergence of industrial socie-

11. J.-P. Sartre, *L'Etre et le Néant*, Gallimard, Paris (1948), part I, chapter II, 2: 'Les conduites de mauvaise foi', 94–5.

12. C. P. Snow, 'The Moral Un-neutrality of Science', in *The New Scientist, Essays on the Methods and Value of Modern Science* (ed. Obler-Estrin), Doubleday, New York (1962), 135.

13. P. L. Kapitza, Address to the Royal Society in honour of Lord Rutherford (17 May 1966): *Nature*, **210**, 783.

14. Don K. Price, *The Scientific Estate*, The Belknap Press of Harvard University (1965), 43.

ties which neither differences of ideology nor differences of political structure prevent from subordinating research activities to the same imperatives of the public authority. 'The imperatives of technology and organisation, not the images of ideology, are what determine the shape of economic society.'[15]

If science instils itself in the heart of politics, it is first and foremost because without the state it cannot meet its own demands. There is no longer any private patron or foundation which can bear the cost of the investments which science needs and which, by definition, are infinite. The change of scale which has occurred since the second world war has placed science in a growing dependence on the public authorities. In twenty years from 1947 to 1967 research and development expenditure in the United States went up fifteenfold, while gross national product merely trebled. Expenditure on fundamental research alone went up more than fivefold, most of it coming from the public purse. The total investment needed by the Manhattan Project, deemed to be the most gigantic research undertaking of all time, amounted to less than one-tenth of the *annual* United States research and development effort today.[16] On the Soviet side, investments in scientific and technical research seem no less than those of the United States, and if the absolute gap between the American effort and that of other industrialised countries, especially the European countries, seems enormous, the relative growth in research expenditure has been no slower on the east of the Atlantic and represents no less heavy a commitment for the state.

The growth of research costs and the increase in the number of researchers are both the effect and the cause of this commitment; they compel the authorities to intervene and at the same time they are the consequence of the major part assumed by the authorities in the support of scientific and technical research. The development of what is called 'big science', of laboratories organised around equipment and research teams whose installations and functioning call for multi-million dollar investments, has been possible only by virtue of and in the light of the intervention of the authorities, involving, moreover, in many instances, the pooling of resources among several nations.[17]

But without science, neither can the public authorities meet their own

15. J. K. Galbraith, *The New Industrial State*, Hamish Hamilton, London (1967), 7.

16. *National Science Policies: the United States*, OECD, Paris (1968), 29–34.

17. See J.-J. Salomon, *International Scientific Organisations*, OECD, Paris (1965), Introduction; and 'International Scientific Policy', *Minerva*, 2, no. 4 (winter 1964).

multiple needs, enlarged or created by the very development of scientific and technical research. The state must take on scientists in the formulation and execution of policy as advisers, administrators, diplomats, strategists. Hence the rise of a new form of technocracy whose influence extends beyond purely technical questions relating to science as such or the exploitation of its results. This is the theme which Lord Snow has illustrated in connection with the part played by Sir Henry Tizard and Lord Cherwell in the decisions taken before and during the second world war on strategic bombing and the conflict between the two men on the priority to be given to radar:

> One of the most bizarre features of any advanced industrial society in our time is that the cardinal choices have to be made by a handful of men: in secret; and, at least in legal form, by men who cannot have a first-hand knowledge of what those choices depend upon or what their results may be. When I say 'advanced industrial society' I am thinking in the first place of the three in which I am most interested—the United States, the Soviet Union and my own country. And when I say the 'cardinal choices' I mean those which determine in the crudest sense whether we live or die.[18]

In the following pages, in analysing the political role of scientists, we shall use the term *technonature* to describe the theatre of this new relation between knowledge and power. The situation of the scientist in the modern world is circumscribed in an area of political decisions which affect his work and which are influenced by his work. Within this area science operates as a technique like any other; *it is the manipulation of natural forces in the light of political decisions,* which at the same time constitutes a source of new problems for the authorities and depends upon the objectives which those authorities are pursuing. Technonature validates the alliance of science, in its specifically scientific character, with ideology as instruments in the hands of power; it designates not a clan, a group or an élite aiming at power for its own sake, but the situation in which science, conceived as the consideration of truth, can no longer be dissociated from the function which it fulfils or from the power which it wields as a political tool. It took an Einstein to give the atomic scientists access to the summit of the United States executive, because in 1939 neither politicians nor researchers were ready to understand each other or to attach any weight to each other. Today, liaison between government and science is institutionalised for questions which are a matter of pure politics, as well as for questions which affect the development of scientific research. The government of science is part of government itself.

18. C. P. Snow, *Science and Government* (1960), reissued by the New American Library (1962), 9.

Technonature is the theatre in which the meaning of science is conceived as purely instrumental, or in other words as being subordinate to values and ends to which scientists claim to be, as of right, strangers. In technonature they find that the apparently neutral procedures of science do not shelter them from the antinomies of 'the ethics of responsibility'. Technonature will not allow science to be regarded as an enterprise indifferent to the values, ideologies and conflicts which throw more than a suspicion of irrationality on to human affairs; while boasting of the rationality of scientific methods and results, it proves, on the contrary, that this suspicion also falls on the affairs of science itself.

III

One of the themes of this book is that, whatever the differences of degree, institutions or procedures, there is no difference in kind between one country and another in the relation of science to power: technonature is a world-wide phenomenon, inherent in the characteristics not of nations or of political régimes, but of the contemporary process of industrialisation. If many of our references are to the United States, it is because this relation assumed the institutionalised form which it now has in all industrialised countries earlier and on a larger scale in the United States, with more acute problems, greater awareness and deeper reflection than elsewhere, though these features can everywhere be traced back to the same source.

We emphasise this point not as a justification for referring to a literature which has proliferated in the United States to the point of becoming a vogue, but in order to deplore the fact that in France similar, if not identical, problems have not aroused the same effort of research and reflection, and, above all, have not induced the universities to take a greater interest in them. There are today nearly a hundred American universities which devote programmes, regular first degree courses or doctoral seminars to these questions; in France the questions are barely beginning to be the subject of university work, mainly in economics, but in such divergent ways from one department to another, even within the same faculty, that one despairs of seeing the organisation of concerted research on the relations between science and society among representatives of several disciplines, as is being done elsewhere (and not only in the United States). If this book calls attention to the importance of the issues at stake and the need to devote genuinely multi-disciplinary research to them, it will already have justified itself as a 'thesis', that is to say, as an attempt at synthesis, inviting subsequent challenge by other scientists.

Sole responsibility for this book lies with the author, but that fact makes him all the freer to thank all those who, in one way or another, have sustained him in his effort. Quickly realising the interest of these questions for the university, Raymond Aron and Georges Canguilhem encouraged a work which was threatened by many professional constraints, and were kind enough to allow it to go forward under the auspices of the Sorbonne. It is impossible to estimate what the text in its final form owes to the attentive and sometimes horrified, but always friendly, reading of André Lwoff.

So many interlocutors in the United States took part in this search for a meaning that I am afraid to name them in case I might leave some out; I would refer here to those who, in the course of constant and repeated talks, contributed more specially to throwing light on my approach, without, of course, necessarily agreeing with the meaning I have finally tried to bring out: David Z. Beckler of the Office of Science and Technology, Robert Gilpin of Princeton University, Norman Kaplan of Northeastern University, Harvey Brooks, Don K. Price, Emmanuel G. Mesthene, Jürgen Schmandt of Harvard University, and Eugene B. Skolnikoff of the Massachussets Institute of Technology.

I could not have completed this book without the leave which Professor Thorkil Kristensen, then Secretary-General of the OECD, and Dr Alexander King, Director-General for Scientific Affairs, kindly granted me to accept the invitation of the MIT Center for International Studies to conduct a seminar there in 1968–9 on science and public affairs in Europe; a leave for which I am all the more grateful to them, since I found in Cambridge exceptional working conditions, teaching colleagues and students from whom I learned a great deal. At a moment when I despaired of coming to the end of this book it was Suzanne Berger, Professor of Political Science at MIT, who showed me, on the roads of Kancamagus, how it could stand on its own.

And, in tribute to his memory, I cannot speak without emotion of what the whole of this book owes to the long discussions I had in Princeton and Paris with J. Robert Oppenheimer, whose drama embodies the price which scientists have paid in becoming the subject matter and the stake of politics.

Paris—Cambridge, Mass.
1968–70

PART I

The New World

The tottering of the world is suggested only by some symptoms here and there: frivolity as well as the boredom that open up in the establishment and the indeterminate apprehension of something unknown are the harbingers of a forthcoming change. This gradual crumbling which did not alter the physiognomy of the whole is interrupted by the break of day that, like lightning, all at once reveals the edifice of the new world.

HEGEL

CHAPTER 1

The Origins

The encounter of science and power is in no way uniquely of our time. As Condorcet wrote:

> In every century Princes have been found to love the sciences and even to cultivate them, to attract Savants to their palaces and to reward by their favours and their amity men who afforded them a sure and constant refuge from world-weariness, a sort of disease to which supreme power seems particularly prone.[1]

And ever since Plato there has been no lack of savants (in those days they were philosophers) who have been more or less happy to act as counsellors to princes, in the hope of inspiring their political action with the underlying principles of their own knowledge, their own philosophy. Antiquity is rich in examples demonstrating the ambiguity and unrest that have always imbued the relations between knowledge and the polity. Nothing is more tempting than analogy; why not go back to the trial of Socrates to account for 'the Oppenheimer case'?[2]

But this type of anology can be drawn only by disregarding precisely what distinguishes modern science from ancient science, that is to say, at the cost of 'forgetting' the major change involved in the 'scientific revolution' of the sixteenth and seventeenth centuries. There were no doubt earlier instances where the political power intervened to act as Maecenas, patron or client, but it was all the more probable that its expectations would not be disappointed, since ancient science shrank from offering practical applications. The example of Archimedes, so often cited, is instructive; it is enough to read Plutarch to realise that the greatest engineer of antiquity was no more interested in turning his engineering work to advantage than was his patron, Hiero of Syracuse:

1. *Eloge des académiciens de l'Académie royale des sciences* (1773 edition), Foreword, 1.
2. The parallel has, moreover, been drawn, not unskilfully; see Sanford A. Lakoff, 'The Trial of Oppenheimer', in *Knowledge and Power*, The Free Press, New York (1967), 65–86.

Archimedes did not regard his military inventions [which were nevertheless used with success against the Romans at Syracuse] as an achievement of any importance, but merely as a by-product, which he occasionally pursued for his own amusement, of his serious work, namely the study of geometry.[3]

In the admirable words of Bachelard:

> So long as the will to power was naïve so long as it was philosophic, so long as it was Nietzschean, it was effective, for good as well as for evil, only on the individual scale. Nietzsche influences his readers; a reader of Nietzsche has only a derisory power of action. But as soon as man effectively masters the powers of matter, as soon as he no longer dreams of intangible elements and hooked atoms, but really organises new bodies and administers real forces, he approaches the will to power fortified by objective verification. He becomes a truthful magician, a daemon of the positive.[4]

This transition from 'naïve will to power' to 'proven will to power' no doubt has its roots in ancient science, but it is only with the advent of modern science that it has been capable of fulfilment. It is as well to recall the main stages of this transformation if we want to understand the changes which we ourselves are still witnessing. We shall merely take a rapid survey, an overview rather than a deliberate inquiry, designed not to throw light on points of history, but rather to delineate the intellectual horizon beyond which the relation between knowledge and the polity takes a new turn.

The scientific revolution of the seventeenth century

Although the whole concept of science policy is very recent, grounded on a close and, above all, irreversible relation between science and the public authorities, this relation made its appearance from the beginning of modern science and, indeed, *with* modern science, as if it were implicit in the new perspective in which the science places its theoretical approach. The mathematisation of nature, whereby in Husserl's words man measures the life-world 'for a well-fitting *garb of ideas*, that of the so-called objectively scientific truths', made possible 'methodical prediction' and 'strict application' to things: the sense of knowledge became action.[5] In contrast with ancient science, the theoretical

3. *Life of Marcellus* (trans. Ian Scott-Kilvert), Penguin, Harmondsworth (1965), 98. Cited in Amyot's French translation by P. M. Schuhl, *Machinisme et Philosophie*, Félix Alcan, Paris (1938), 15.

4. G. Bachelard, *Le Matérialisme rationnel*, P.U.F., Paris (1953), 5.

5. E. Husserl, *The Crisis of European Sciences and Transcendental Phenomenology* (trans. David Carr), Northwestern University Press, Evanston, Illinois (1970), 51.

outlook of modern science marked a change of direction in attitudes towards things. At the same time, by its promises of application—even if for a long while it was far from able to keep them—knowledge entered a new relationship with political power; the political take-off of modern science was inherent in its intellectual take-off.

Science aims at extending our knowledge and understanding of nature. This hackneyed definition is questionable only if we rely on it to make science an activity detached from the concerns of the world, purely inward-looking, owing nothing to other human activities and in no way serving them. The idea of a science which will not let itself be useful to society, which compels the researcher to immure his interests and his curiosity in the famous ivory tower, goes back to the days when science had no sway, and above all sought no sway, over nature, that is to say, to Greek antiquity and the Middle Ages. In other words, it goes back to the days when science, at least as we understand it today, did not exist. By an irony of history, those who refuse science any social function do not see that what they are talking about is nothing like what they know or practise today; for the science which is called modern, which imposed itself after the 'Galilean revolution', set itself precisely *against* the concept of a science which divorced theory from practice.

Even in the seventeenth century, university teachers, the heirs and disciples of Aristotle, were not only strangers to the idea of experimental science, they were downright hostile to it. In their eyes, true science was based not on the interaction of experience and reason, reason being tested by contact with experience, but on the discovery by unaided reason of the principles and truths which experience itself prevented us from attaining, like a veil thrown between appearance and reality. The true science was *contemplation*, that is to say, a purely intellectual vision of realities beyond the sensible world. If the 'official' savants had no laboratory in which to test their conception of the real by manipulation and experiment, it is not because they did not know how to use one, but because they did not want one, for fear of failure. No period demonstrates better how the state of science is linked to a whole conception of the world; this science, which was contemplation, was reserved for 'free men' doing 'liberal' work, while technology was the preserve of artisans whose work was 'servile'; just as technology was lower than science, the artisan was lower than the 'free man' who was the savant. It cannot be too greatly emphasised how heavily this social prejudice weighed upon science in those days; in any event, it explains the scorn or disdain with which the Middle Ages treated the work of the rare savants of antiquity who dreamed of a mechanistic science which would influence nature in order to transform it.[6]

This same prejudice has outlasted the seventeenth century in the idea, still so flourishing in our own time, of a radical distinction between fundamental research and applied research, between science and technology, between the pure scientist and the engineer. We may recall the passage in which Plato, after recognising the outstanding services which the engineer renders to the city by contributing to its defence, went on to say:

> Nevertheless, you despise him and his art, and sneeringly call him an engine-maker, and you will not allow your daughters to marry his son, or marry your son to his daughters.[7]

Even though this scorn was based on an obsolete scale of values and social functions, it is still found today in the prestige enjoyed by the 'pure' researcher in contrast with the engineer. No doubt science and technology did not really begin to go hand in hand and to depend on each other until quite recently, starting in the nineteenth century. For a long time invention, especially industrial invention, was in no way aided by scientific discovery: rather, if one thinks of scientific instruments, science seems more indebted to technology than *vice versa*.[8] It is not, however, because modern science, born of Galileo, Descartes and Newton, took a long time to fulfil its promises of rapid application that we can disregard the precise feature which most distinguishes modern from ancient science, namely its determination and ambition to use knowledge as a link between *theoria* and *techne*.

The idea of an alliance between science and society springs from the whole nature of modern science. If art, in Bacon's words, is man added to nature, the artisan is man linked to power; as science draws nearer to the workshop, it tends to draw nearer to the political authorities to whom it offers the use of its artifices. From the start, political authorities have asked savants to tackle problems of special concern to them; from the start, the savants have turned to the authorities, pleading the utility of possible results to justify support for what would nowadays be classified as pure research. Let us listen to Descartes:

6. See, in particular, P. M. Schuhl, *op. cit.*, note 3 above, and Robert Lenoble, 'Origines de la pensée scientifique moderne: l'héritage médiéval', in *Histoire de la Science*, Pléiade, Gallimard, Paris (1957), 375–8.

7. *Gorgias* (trans. Jowett), 512c, Oxford University Press (1920).

8. See, in particular, Maurice Daumas, *Les Instruments scientifiques aux XVIIe et XVIIIe siècles*, P.U.F., Paris (1953).

My Lord Cardinal (Richelieu) should have left you two or three of his millions to make all the experiments necessary to discover the special nature of each body, and I doubt not that great knowledge could have been gained which would have benefited the public far more than all the victories which can be won by making war.[9]

The utility of knowledge: the formula is the same as the famous words in the *Discours de la Méthode*, in which Descartes envisages a 'practical philosophy which should replace the speculative philosophy which is taught in the schools' and which should make men 'as it were the masters and possessors of nature'.[10] A science which holds itself out as useful needs institutions capable of supporting its efforts. The vision which looks upon nature as a machine demands machines to prove itself by experience: instruments, privileged places, constant exchanges—and therefore money and organisation.

It is significant that modern science owes to the same man both the principles of its method and the (imagined) bases of its organisation. In the utopia of the *New Atlantis*, published in 1627, a year after his death, Francis Bacon described an imaginary voyage to a South Sea island and the discovery of an entirely different civilisation, resembling the utopias of Campanella and More. But whereas these utopias discovered the conventional landscape of a better society on condition of escaping from the corruptions of the day and returning to the supposed paradise of the primitive community, Bacon's utopia imagined a society organised on the principle of scientific research. His vision substituted the exploration of the promised land for the nostalgia for paradise lost:

> The end of our foundation is the knowledge of causes, and secret motions of things; and the enlarging of the bounds of human empire, to the effecting of all things possible.[11]

The vision also substituted, in the social hierarchy, the prestige of the researcher for that of the great ones of the earth; the return to the island of one of the 'fathers' of the house was accompanied by honours and ceremonies normally reserved for a monarch or the head of a state or a church.[12] Having postulated the recognition of science as a public service, Bacon endowed it with an institution in which everything was organised for the production of discovery, according to a division of

9. Letter to Mersenne, 4 January 1643 (ed. Adam–Tannery), vol. 3, 610.
10. 'For they [the general concepts of physics] have shown me that it is possible to attain knowledge which is very useful in life. . . .' *Discours de la méthode*, Pléiade, Gallimard, Paris (1952), 168.
11. *The New Atlantis* (ed. A. B. Gough), Oxford (1924), 135.
12. *Ibid.*, 233.

labour which, step by step, strictly obeyed the rules of his method. From the 'merchants of light' collecting knowledge from foreign countries, to the 'depredators' collecting knowledge from books, the 'pioneers' trying new experiments, and the 'interpreters of nature' turning experiment into action, 'Salomon's House' fulfilled most of the functions which in our day correspond to the development of research, and to the needs of the state: scientific attachés, information services, administrators and even meetings of advisory committees to assess the results obtained, to programme research and to decide which inventions must be kept on the secret list.

It is even more significant that the *New Atlantis*, a utopia in the Renaissance sense, a 'model' in the modern sense, was the work not of a savant but of a politician: barrister, Privy Councillor, Attorney-General, Lord Chancellor and philosopher in his leisure moments, Bacon was never a practising scientist. William Harvey derided his scientific attainments and said that 'he wrote philosophy like a Lord Chancellor'. But it was precisely as a statesman, conscious of the links between experimental research and society, that he dreamed of a new social order in which organised and planned science would be systematically developed; in the history of the relations between science and politics, he is the first statesman of science.[13] It is alleged, said Bacon, that learning softens men's minds and makes them less fit for the exercise of arms, that it perverts their disposition for matters of government and policy by making them too curious and irresolute, that it diverts them from action and business and gives them a love of leisure and privacy. But, he rejoined, Alexander and Julius Caesar were great scholars as well as great generals; philosophers are useful, and even necessary to the state; men of learning work harder than others, because they take pleasure in the work and do not do it for profit; and finally, the state should benefit rather than suffer from 'duty taught and understood' instead of a blind custom of obedience.[14] For Bacon, the greatest justification of science was its utility, since it was reflected in power. Not that we should turn our backs on the theoretical activity on which the whole science of the ancients was founded; the task was rather to reconcile theory and action. It is true that Bacon emphasised the technological resources of learning, which made it 'the servitor and interpreter of nature',[15] where Descartes emphasised the speculative

13. J. G. Crowther, *Francis Bacon, The First Statesman of Science*, The Cresset Press, London (1960).
14. *Advancement of Learning* (ed. G. W. Kitchin), London (1950), 13–14.
15. *Instauratio Magna, in fine*.

power of reason to dominate nature. Between the two of them, the idea of science differed in the light of a different conception of nature: Bacon's concept still referred back to hidden powers whose secrets must be snatched from them; Descartes' concept banished mysteries by imposing geometrical order on nature. But the science envisaged by Bacon was not thereby cut down to the application of the practical arts which his thinking, in contrast to the mediaeval attitude, regarded as legitimate.

The intellectual New World which he opened up, and of which he fully realised he was the Christopher Columbus—the frontispiece of the *Instauratio Magna* is an engraving of a caravel sailing through the Pillars of Hercules, which to the ancients were the end of the known world—was that in which science imposes itself both by its instrumental character and by its speculative character. Today, the very existence of organised science policies indicates that science is nowadays regarded purely from its instrumental aspect. For Bacon this aspect was inseparable from the other; the New World must achieve the conjunction of contemplation and action which alone could conquer nature and enslave it to the ends of man: *scientia et potentia humana in idem coincidunt.* That there has been a misuse of the fruits of science is another matter. Francis Bacon, Lord Verulam, noted in his journal that he had no greater ambition than to serve the whole of mankind by science; he contrasted the glory of the statesman with that of the inventor, which did not burden the conscience with remorse.[16] Three centuries later, the conquest of the atom was in turn to open up a New World, where the glory of the inventor was to weigh no less heavily on the collective conscience than that of the statesman.

The utopia of the *New Atlantis* is the idea of a constituted body which would take its place in the state alongside the institutions responsible for justice, the army, education and finance, and whose sole service would be scientific research. Bacon may have been inspired by the assemblies of men of letters or of artists, and sometimes of alchemists, which met together from the sixteenth century onwards to deal more or less periodically with scientific subjects. As early as 1560, G. B. della Porta had founded an academy in Naples to which admission could be gained only by contributing a secret discovery. But it was in Rome in 1603 that there appeared the first scientific academy in the sense of an institution set up in the shadow and with the support of a political

16. *Cogitata et Visa, The Works of Francis Bacon* (ed. Spedding-Ellis), second edition, London (1887–1892), vol. 3, 610–11.

authority for the purposes of research in the natural sciences; the Accademia dei Lincei was founded by Prince Federico Cesi, Duke of Acqua-Sparta, who housed it in his palace and endowed it with a botanical garden, a natural history collection and a library. The Lincei wore a gold ring, set with an emerald engraved with a lynx, the badge of the Prince and of the Academy; they were still more like members of a secret society than of a public institution and their status depended on the fortune of the Prince. The Accademia dei Lincei had no more permanent status than the Accademia del Cimento founded a few years later at Florence, whose title—experiment—was a whole programme in itself. On the death of its patron the Accademia dei Lincei ceased to exist.

The creation of academies with official status was the culmination of a long evolution of society which began before the Renaissance. The sixteenth century saw the formation of a rich bourgeoisie who took an interest in the arts of the engineer for which the 'official' science of the day had nothing but scorn; modern science was born on the fringes of the university and often in opposition to it. Thus, in Italy, the development of commerce and trade led to the development of the fine arts and at the same time gave a new value to trades which had so far been regarded as mercenary. It was not only the burghers, such as Vincent Galileo, the father of the great physicist, who gained entry to intellectual life, but technicians too, whose competence and range of knowledge soon entitled them to set up as rivals of the university. A link was formed between the function of the craftsman and that of the savant which battered a breach in the Aristotelian principle of the distinction between science, the contemplation of eternal truths, and the arts, the empirical manipulation of appearances. Physics gradually freed itself from metaphysical speculation and enlisted in the service of man, because outlooks and manners were changing, at the same time that the spirit of science itself was changing. While the university was still the prisoner of Aristotle's disciples, teaching a science which had no contact with reality, new research was developing in the shadow of the princes and the great burghers. To start with, this research was the work of a few isolated individuals whom the princes sought to monopolise in their own interests and who met together only on the princes' initiative. Quite different were the beginnings of the Royal Society and the Académie des Sciences of Paris; both sprang from private meetings to which royal patronage was accorded only later. The founders of the Royal Society met in private houses or taverns, or sometimes at Gre-

sham College, long before they were recognised by Royal Charter in
1662, just as the founders of the Académie des Sciences met for a long
time in Thévenot's house before being accorded official status, thanks to
Colbert, in 1666.[17] But it took less than half a century to progress from
Baconian utopia to reality; the advancement of learning and the institutionalisation of research, of which Bacon was the ardent promoter,
were officially linked to political power. Undoubtedly this link took
different forms in the two institutions; the support accorded to the
Royal Society was purely formal and its resources, drawn as best they
might be from members' contributions and gifts, verged on poverty (up
to 1740 its total annual income did not exceed £232); the protection
which Louis XIV accorded to the Académie des Sciences guaranteed its
financial support (the costs of experiments and instruments were paid
and its members received an income) but, at the same time, implied
dependence; after Colbert's death the decline of the Académie was to
be the price of a royal favour which was too self-interested, 'curious
research or research which is, as it were, an amusement', having to yield
place, according to the injunction of Louvois himself, to 'useful research which may relate to the service of the King and the State'.[18]

In fact the Royal Society aimed at the useful no less than the Académie des Sciences, since the Gresham College amateurs won the
patronage of Charles II only because of the interest they showed in the
problems raised by trade, manufacturing and shipping. Robert K.
Merton has very well described how

> from the outset the members of the Royal Society were concerned to justify
> their activities (to the Court and the lay public) and tried to show practical
> results of their work as soon as they could.[19]

It was, moreover, the example of shipping which enabled him to show
that its utilitarian motives had very soon not only influenced the direction of certain work, but had even led to certain discoveries which might
rather have been expected to be the fruit of pure research; an argument
which is willingly inverted nowadays to recall that theoretical research

17. See Martha Ornstein, *The Role of the Scientific Societies in the Seventeenth Century*, Chicago U.P. (1928), and René Taton, *Les origines de l'Académie royale des Sciences*, address at the Palais de la Découverte, Paris (1965).
18. Joseph Bertrand, *L'Académie des sciences et les académiciens de 1666 à
1793*, Paris (1869), 40, cited by Ornstein, *op. cit.*, note 17 above, 157.
19. Robert K. Merton, 'Science and Economy of Seventeenth-Century England', in *The Sociology of Science* (ed. Barber, Hirsch), The Free Press, New
York (1962), 67–88, and above all, his PhD dissertation, *Science, Technology
and Society in 17th Century England*, Bruges, Imprimerie Sainte-Catherine
(1938; reissued by Harper Torchbooks, New York, 1970).

the most remote from any utilitarian concern is often rich in practical applications.

The calculation of longitude at sea appeared an essential trump card in the mastery of the seas; the perfection of a rigorous method was so vital for England, whose island position raised problems which did not escape the notice of the scientists of the day,[20] that a special Act, passed in 1714, on the recommendation of Newton, offered a reward to anyone who succeeded.

> The importance attached to solving this problem can be judged by all the rewards offered by other governments. The Dutch government tried to persuade Galileo to apply his talents to finding a solution; Philip III of Spain also offered a reward, and in 1716 the Regent Philippe of Orléans established a prize of 100,000 francs.[21]

The Charter of the Royal Society, adopted in 1663, set it the object of

> perfecting the knowledge of natural things and of all useful arts, manufactures, mechanical practices, engines and inventions by experiment, not meddling with Divinity, Metaphysics, Moralls, Politicks, Grammar, Rhetorick or Logick.

Politics was excluded in the sense of opinion or party allegiance which might interfere with the scientific approach, but in another sense it was implicit as a conjunction of two interests, the service of science and the service of the state: the apoliticism of the intellectual approach did not rule out a contract with power.

> It cannot be sufficiently stressed that it was the experimental character of science which encouraged the creation of scientific societies.[22]

This same character involved the constantly closer liaison of the state and research in a convergence of interests that was to define precisely the domain of science policy.

In 1752, Maupertuis, presiding over the Academy of Berlin, submitted to Frederick II a detailed programme of investigation.

> I merely desire here to turn your eyes towards certain research useful for mankind, curious for the learned and which the present state of science seems to place within reach of our senses. . . .[23]

20. For example, Edmund Halley's report to the Royal Society: 'That the Inhabitants of an Island, or of any State that would defend an Island, must be masters of the sea and superior in naval force to any neighbour that shall think fit to attack it, is what I suppose needs no argument to enforce.' Cited in Merton, 'Science and Economy of Seventeenth Century England', 87, note 69.

21. *Eléments de théorie et de méthode sociologique*, 495, note 40.

22. Ornstein, *op. cit.*, 67.

23. *Lettres sur le progrès des sciences, Oeuvres de Maupertuis* (1752 edition), Dresden, 3.

To satisfy both the curiosity of the learned and the needs of mankind, it was necessary for the authorities to step in to help with research that called for resources beyond the means of individuals.

> There are sciences over which the will of kings has no immediate influence; it can procure advancement there only in so far as the advantages which it attaches to their study can multiply the number and the efforts of those who apply themselves to them. But there are other sciences which for their progress urgently need the power of sovereigns; they are all those which require greater expenditure than individuals can make or experiments which would not ordinarily be practicable.[24]

The advancement of knowledge henceforth passed by way of political power; at the same time, the interest of the state lay in the consultation of savants. The logic of this reciprocal relation was very clear to Condorcet: science instituted as a body of the state needed 'the enlightened protection of the government', just as the government, if it were to benefit from science, needed the counsels of 'a learned company'. The ideas of choice and priority in the allocation of resources were already emerging.

> Projects are constantly being presented to the government, always confidently announced as calculated singularly to extend or to perfect the most useful arts. It would be equally dangerous to adopt these projects without examination or lightly to renounce the advantages they promise. We therefore need a Society of learned men which, judging without prejudice and remote from any special interest, will enlighten the government on the means proposed to it and will show what is the precise degree of utility of those which should be adopted and how far they can be expected to succeed.[25]

The political revolution of the eighteenth century

It is once again to Condorcet, the last of the *philosophes* and the only one who took part in the French Revolution, that we must refer for the next stage, in which we find the idea of progress inspiring a new relation between science and political power. Bacon's utopia tended to recognise the legitimacy of modern science as an instrument both of knowledge and of action; the legitimacy of the method and the legitimacy of the institution both stemmed from the same combat against ancient science. But it was only for the scientists that the New Atlantis was a promised land; the scientist was king, but the system was none the less a monarchy. Pursuing Bacon's utopia, but in the spirit of the century which

24. *Ibid.*, 6–7.
25. *Eloges des académiciens de l'Académie royale des sciences* (1773 edition), 2–3.

looked upon the progress of knowledge as the instrument of social and political progress, Condorcet visualised a scientific society based on the democratic principle, whose establishment represented the culmination of the historical development of mankind.

In the *Fragment sur l'Atlantide*, which he wrote on the eve of his death as an act of faith in the power of reason to ensure the happiness and equality of men, he combined Bacon's trust in organised science with the revolutionary trust in science as an organising principle, to reorganise the social system as a whole. The century of the Enlightenment broadened the technical utopia into a messianic vision; just as the reform of the body politic depended upon the advancement of knowledge, so the scientific institution provided the model of civil society as it would be perfected in the final stage, 'the tenth stage', of historical development: 'the general meeting of the world's savants in a universal republic of sciences, the only one whose project and utility are not a childish illusion'.[26]

Thus the relation between power and knowledge was transformed; from an isolated island, as it were exiled from society as a whole in the Baconian spirit, the society of savants was called upon to occupy a privileged place, to assume the role of a prototype of the ideal society, itself constituting the model of enlightened power as the most accomplished instrument of historical destiny:

> I have set myself in a truly free country where real equality reigns, where the general enlightenment and the knowledge of the rights of man leave no room for fear that the foundation of the public happiness will ever be sought in equality of ignorance and folly.[27]

A king could perhaps have ensured, from personal whim,

> that here men would attempt and suffer as much to arrive at the knowledge of a few truths as they would for the love of gold, and that there instruments whose cost is beyond the reach of any private fortune would snatch some of its secrets from the sky,[28]

whereas the triumph of democracy ensured continuous and systematic support for science:

> this constancy, this set of views embracing a long sequence of generations, extending to the whole system of sciences, is something which can be expected only from a people whose laws have been dictated and whose institutions have been combined by a strong and pure reason.[29]

26. *Esquisse d'un tableau historique des progrès de l'esprit humain*, Part Two, *Oeuvres complètes* (ed. Garat-Cabanis), vol. 8, Paris (1804), 521.
27. *Ibid.*, 522.   28. *Ibid.*, 516.   29. *Ibid.*, 517.

In so far as it was to be acquired under the auspices of democracy, this support ruled out any government interference which was not in line with the needs of science; the hypothesis of liberty attained enabled us to set limits to the intervention of the state, whose representatives must be 'enlightened enough to feel that they must not direct the work but second it, that they must not regulate the discoveries but profit from them'.[30] The line of demarcation was drawn, by definition, by the equilibrium of the whole social system, which now obeyed only the voice of reason.

> Thus, the parts of the social system apparently most remote from each other come together at certain points. Thus, if reason is to exert its full empire over one part it must succeed in spreading itself over all parts.[31]

In the 'tenth age' of human history, when science would triumph over superstitions and inequalities, the problem of the allocation of resources would be no more than a false problem, since the society of savants must be aware of 'the whole dignity and independence which the individual possesses in a free people'.[32] Neither give nor take: state policy towards science was all the easier to determine since its proper principle was reason:

> It is for the association alone to judge independently what it believes should be undertaken to speed up the progress of the sciences. It is for the public authorities to judge, with the same independence, which of its projects merit support, or even munificence. Let us not despair of seeing the time come when this division can be effected through reason alone. . . .[33]

It is true that Condorcet was not thinking of works so costly that they needed massive state support; his imagined scientific society would live mainly on the contributions and gifts of its members, with a reserve fund whose income would suffice for the greatest needs.

> The public authorities could be expected and desired to provide either facilities for correspondence or means of distant travel or the grant of sites, some of which might prove strictly necessary, and the good offices of its agents at home and abroad, which might be essential for certain research.[34]

The state would partake in the fruits of research, but would not be its primary customer, and therefore not its primary source of finance. The progress of the sciences was to be aimed at as a social goal because it must necessarily affect the progress of society; that did not mean that science would lose its autonomy in order to meet the short-term objectives of the state.

30. *Ibid.*, 598. 31. *Ibid.*, 596. 32. *Ibid.*, 597.
33. *Ibid.*, 598. 34. *Ibid.*, 597.

The eighteenth century may seem the expression of a dreamed-of alliance between science and politics—dreamed of as the hope of a world which looked to the benefits of science to solve the problems of the relations between the individual and the state, but also dreamed of as the mirage of a state of science which could still do no more than promise, in the absence of rapid applications. The state came into Condorcet's scheme of things only as the financial guarantor of the development of research; science policy was reduced to policy *for* science, which contributed to social progress more by its purpose than by its applications. The instrumental character of modern science, stressed by Bacon, was no longer perceived except in its ideological context; science was power, but primarily power to emancipate the mind. If it acted on nature, it was to liberate social nature; it served the state through the enlightenment it spread, but the state could not bring pressure to bear on it to orient its efforts. The society of savants was all the better protected against demands from power alien to its own interests, since it was the model of power. For Condorcet, in sum, it was the state which was the servant of science.

The experience of the French Revolution

The revolution of which Condorcet was one of the victims aspired to the same end as that which inspired the *Fragment sur l'Atlantide*, namely the liberation of minds and people. But under the revolution the relation between science and the state would not be a one-way relation as Condorcet conceived it. In practice, support of science by the state would depend on support of the state by science. The summing up of the contribution of science to the defence of the Republic, presented by Fourcroy on 3 January 1795, was prefaced by a declaration which Condorcet himself would no doubt not have disavowed:

> The Committee felt that it should take this opportunity of lightly sketching for the National Convention the discoveries which have rendered such good service to the cause of liberty and of making known to Europe how greatly the sciences and arts, cultivated and perfected by a free people, influence the wisdom and success of its government.[35]

If for the first time in history science was found playing a direct part in the march of state affairs, the reason is precisely that it was the density of the state which was at stake. As Fourcroy says in another

---

35. *Rapport sur les arts qui ont servi à la défense de la République*, cited by J. Fayet, *La Révolution Française et la Science*, Marcel Rivière, Paris (1960), 226.

report: 'The War has become a happy occasion for the French Republic to develop the full power of the arts and to exercise the genius of
savants.'[36] The 'happy occasion' was born not of the aspirations of
science, but of the threats hanging over the revolution. In associating
with itself from the second day of its creation, 'a Commission of four
citizens learned in chemistry and mechanics specially responsible for
seeking out and proving new means of defence',[37] the Committee of
Public Safety launched the first measures of 'direction of manpower'
which were to lead to 'the mobilisation of savants in the year II'. This
'Congrès des Savants',[38] at the service of the Committee of Public
Safety, constituted the first example in history of a government relying
on the counsels of a group of scientists, giving it technical missions of
organisation and development and associating it by virtue of its scientific qualifications with the definition and execution of its policy. It
introduced the era of policy *through* science, the originating motives for
which would *always* be military.

The number of savants who contributed to the national defence was
large, and they were by no means the least distinguished—Lavoisier,
Carnot, Monge, Berthollet, Chaptal and others. But this was still a long
way from 'the universal republic of sciences' dreamed of by Condorcet,
which would know no disqualification except failure to respect unani-
mously recognised truths. The Republic chose its partisans in the same
way that it chose its victims. Lavoisier was condemned because, among
other things, he embodied the 'learned aristocracy' of the Ancien
Régime, and the gifted men whose support the Committee of Public
Safety ignored or rejected were no fewer than those who were 'mobilised'—Coulomb, Parmentier, Legendre, Lalande, Laplace and others.
Savants were not deemed fit to serve the state merely by virtue of being
savants; they must also give proof of allegiance, they must show not only
that they were thinkers, but that they were orthodox thinkers. 'Happily,'
said Carnot quite bluntly, 'the men from whom we sought enlightenment could not be challenged on the grounds of their probity and civic
spirit any more than on the grounds of their talents and zeal. . . .'[39] For
the first time, a cleavage was introduced into scientific society; revolutionary messianism *politicised* science and, at the same stroke, divided

36. Cited in Fayet, *ibid.*, 237.
37. See A. Mathiez, 'La Mobilisation des savants en l'an II', in *Revue de
Paris* (1 December 1917).
38. Fourcy, *Histoire de l'Ecole polytechnique*, Paris (1828), cited in Fayet,
op. cit., note 35 above, 238.
39. Speech to the Convention (3 November 1793).

it. The reasons why Oppenheimer lost his security clearance in 1954 did
not greatly differ from those which sent Lavoisier to the scaffold;
neither was guilty of 'treason' but both were suspect, Lavoisier because
of his associations with the Farmers-General (whom he had left before
1789) and Oppenheimer because of his contacts with Communist Party
members (which he had broken off before the declaration of war).

'The Republic has no need of savants': it is not certain that these
words, attritubted to Coffinhal, were ever spoken, but, as Joseph Fayet
has shown, 'similar, and sometimes even identical, words were used in
the Assembly or in Committees and reprinted in the journals'.[40] Those
who used them were thinking just as much of the 'privileged' academic
savants as of the 'speculative', 'erudite' or 'witty' savants, for it was the
orientation of their work that was challenged as much as their social
status. In the debate which now opened on the organisation of public
education, the aim was not only to prevent the renewal of 'learned aris-
tocracies' but also to orient science towards concrete utilitarian tasks.
In fact, the Republic did need savants, but they must also produce their
recognised credentials of republicanism, they must turn their backs on
'lofty speculations'[41] and tackle practical, day-to-day problems.

> Free nations [exclaimed Bouquier in the Convention] do not need a caste
> of speculative savants whose minds are constantly wandering along the
> by-ways in the realms of dream and chimera. The sciences of pure specula-
> tion isolate their followers from society and, in the long run, become a
> poison which saps, enervates and destroys republics.[42]

In becoming politicised, the society of savants became split on two
planes, according to the line researchers took in their opinions and the
line they took in their work. The displacement of revolutionary messianism by national messianism was enough for the same cleavage to
divide researchers within their own country and between countries.

Here again, we are a long way from the republic of sciences dreamed
of by Condorcet, which could deal high-handedly with the requirements of the political authorities for the very reason that it was itself the
model of power. The revolution rather tended to show that the closer
the relation between science and the state, the more the state looked to
science to yield short-term benefits. The advantage, in this sense, lay not

40. Fayet, *op. cit.*, 197.
41. Speech of Rudel to the Convention, cited in Fayet, *op. cit.*, 198. See also
the speech by Durand-Maillane: 'The French people, in order to be happy,
need no more science than they must have to attain virtue . . .', *ibid.*
42. *Rapport et Project de décret formant un plan général d'instruction pub-
lique* (9 December 1793), cited in Fayet, *op. cit.*, 199.

with 'pure' research but with applied research, and even more markedly with what is nowadays called development, that is to say, the last stage of research before production. The record of the 'Congrès des Savants' attached to the Committee of Public Safety was not one of discoveries and inventions destined to enrich science; it was a technical record, centred entirely on war provisions, the production of powder and and the manufacture of arms. When inventions with a great future—the telegraph, or balloons—were supported, it was always because of their immediate value; but research on the steam engine was dropped because it yielded no quick advantage. The job was certainly well done and fully attained its ends, but scientific concerns played only a minor part.

It was after the event, and indirectly—by what we should nowadays call 'spin-off'—that this mobilisation of savants was to contribute to the expansion of science. The war was not only a 'happy occasion' for the Republic, it was also a happy occasion for the savants as a social group becoming self-aware and vested with a new power. They enjoyed, said J. B. Biot,

> unlimited credit. It was well known that the Republic owed them its salvation and its existence. They profited from this instant of favour to guarantee France that higher degree of enlightenment which had given it the victory over its enemies.[43]

From their close association with power, not only had they earned the right to impose scientific reorganisation on the country but, through them, science itself was recognised as a social force, that is to say, as an institution which society could no longer do without. There is no better illustration of this change than the creation of the *Ecole Polytechnique*. The 'professionalisation' of scientists had started before the French Revolution and the kind of teaching which the new institutions were to make general certainly had its tradition, but this kind of teaching had no recognised home and, above all, was not systematically directed towards 'higher education'.[44] For the first time, science was to be taught in its own right, at high level, theoretically and practically, at an insti-

43. *Essai sur l'Histoire générale des sciences pendant la Révolution française*, Paris (1803), 58.
44. On the professionalisation of scientists, see Joseph Ben-David, 'The Scientific Role: The Conditions of its Establishment in Europe', *Minerva*, 4, no. 1 (autumn 1965), 15–54. On the genesis of the new institutions founded by the French Revolution and the inheritance they received, see *Enseignement et Diffusion des sciences en France au XVIIIe siècle* (ed. René Taton), Hermann, Paris (1964), in particular René Taton, 'L'Ecole royale du Génie de Mézières', 559–615.

S.P.—2*

tution of higher education equipped with laboratories, fully recognised
and established as a research centre; designed to train engineers and
army officers, the Ecole Polytechnique was to function wholly as a nursery for scientists.

When the dangers which the mobilisation of researchers was intended
to meet became less pressing, the relation between science and politics
ceased to be one-way; it was the scientists' turn to take advantage of
politics. On this point, too, the revolution marked an innovation, since
it inaugurated the unwritten law that the war effort in which the scientists had taken part should be complemented, after the end of hostilities,
by an educational effort. 'Legislators, this is your work!' exclaimed
Grégoire on 31 August 1794, when he presented the Convention with a
list of the work done and the scientific foundations created since the
beginning of the revolution;[45] a few days later, in paying tribute to the
scientists, he stressed their prior claims on the state.

> If the road of liberty lies open before us, it is they who have been the
> pioneers, the legislators of the principles which you have applied.[46]

The wheel had come full circle; in the alliance between science and
politics it is already impossible to tell which of the two was really making use of the other.

The last days of laissez-faire

With the Industrial Revolution, history seems to have accelerated,
according to Auguste Comte's dictum, in harmony with the growing
accumulation of knowledge and power, an accumulation which itself
obeyed a law of gathering speed. The reality of growth production conceived as—having no defined limits—took over from the idea of progress in harmonising rationality and power. But if the process of
industrialisation has appeared to be the inescapable destiny of man, we
still have to postulate the providential nature of history (or more simply
take an optimistic view of the future) to expect power to place itself at
the service of rationality. Auguste Comte was not the first philosopher
to believe that scientists would one day assume power, but he was certainly the last to have any good grounds for this belief.

The new science founded by Comte's *Cours de Philosophie Positive*
took note of the Industrial Revolution as the outstanding event of
modern times: theological thinking was dead, and with it the military

45. *Rapport sur la déstruction operée par le vandalisme.*
46. *Rapport sur les encouragements et récompenses à accorder aux savants,
gens de lettres et aux artistes* (September 1794).

society which had been its temporal expression; the scientists must take over from the priests, and the industrialists and bankers must take over from the warriors. The irrefutable logic of the law of the three estates and the classification of sciences determined the future of mankind, in which scientists were destined to exercise a spiritual mastery while industrialists applied themselves to the rational exploitation of nature. In this division of power, spiritual and temporal, the vocation of the new priests was to moderate the authority, the arbitrariness or the avidity of the new holders of power since, as scientists, they possessed the moral prerogatives which allowed them to check the exercise of temporal power. Even if the political order were still based on force, which always depends on numbers and wealth,[47] Comte's providentialism asserted the subordination of political power to science.

The history of the relations between science and politics from the emergence of the Industrial Revolution has in fact been quite the reverse. In the nineteenth century, far from holding sway over temporal power, scientists never ceased campaigning for recognition of the need for an official policy of aid to scientific research.[48] In most countries, reaction to the French Revolution first inspired pleas for state intervention, until displaced by the success of the German universities and industries in the second half of the century. Whatever these pleas, and wherever they were heard, the argument most commonly put forward was that of the potential utility of scientific discoveries. Thus Charles Babbage, in his pamphlet on the decline of science in England, published in 1830, pointed to the French example to show that the lack of interest shown by the authorities endangered the expansion of industry.[49] The argument was to strike home, since British institutions of higher education were gradually modernised, but these reforms were much more the result of private initiative (in particular the Association

47. In the temporal sphere, according to Comte, it is always the law of the strongest which predominates: 'Such power, in truth, takes an essentially material form, since it always results from greatness and wealth. But it is important to recognise that the social order can never have any other immediate basis. Hobbes's famous principle of the spontaneous domination of force constitutes basically the only major step forward, from Aristotle's time to my own, taken by the positive theory of government.' *Système de politique positive* (ed. by the Société Positiviste), vol. 2, 299.
48. See, in particular, R. Taton, 'Les Conditions du progrès scientifique en Europe occidentale', *Histoire générale des sciences*, part 3, vol. 1, 614–29, P.U.F., Paris (1961).
49. *Reflections on the Decline of Science and on Some of its Causes*, London (1830).

for the Advancement of Science, created by Babbage's friends the following year) than of any public action.

It was another friend of Babbage's, Lieutenant-Colonel Strange,
FRS, who as early as 1872 was to propose the creation of a Ministry
of Science, the first plan for a political authority that would recognise
science as 'an affair of enormous national importance'.[50] There was no
sequel to the project, which aroused as much hostility in scientific
circles as in political circles. Among the scientists, the idea of a centralised body responsible for scientific affairs, even for the sake of providing
greater support, immediately awakened the fear of undue intervention
by the authorities in the direction of research; while among politicians,
the response was that unlikes should not be combined—Lord Derby,
for example, said that the conduct of administrative affairs and the
management of men were something quite different from scientific
research.[51]

Some years later Renan published *l'Avenir de la Science* (he had in
fact been thinking about it since 1848), which contained one of the
most fervent pleas in favour of public support for scientific research,
but this time the argument relied on was no longer the potential use of
discoveries. It was the intrinsic worth of science itself which demanded
public support, an absolute duty, a categorical imperative resulting from
the fact that society was responsible for 'the conditions for the perfecting
of man' and that, among these conditions, science had ousted religion
from the first place:

> From the moment that man's religious instinct comes to be exercised in
> purely scientific and rational form everything which the state used to do
> for the exercise of religion becomes due, as of right, to science, the only
> definitive religion. There will no longer be a budget for religious worship,
> there will be a budget for science, a budget for the arts.[52]

The duty of the state stemmed from the collective investments which
science demanded:

> Individuals cannot build themselves laboratories, create libraries, found
> big scientific establishments. The state therefore *owes* science observatories,

50. J. G. Crowther, *Statesmen of Science*, Cresset Press, London (1952), 254.
51. *Ibid.*, 261. During the discussion of the project, one member of the
Royal Society, protesting against the insufficient support given by the government, said the Treasury was behaving like a man who discharges all his responsibility towards the less fortunate classes by giving a coin to the first beggar
he meets (*ibid.*, 260).
52. Renan, *L'Avenir de la Science*, Calmann-Lévy (1890), 252–3. There are
still countries such as Sweden and Norway where the same man is Minister for
Science and for Religion.

libraries, and scientific establishments. Individuals cannot undertake and publish certain work on their own. The state *owes* them subsidies. Some branches of science (and those the most important) cannot afford their practitioners a living; the state therefore must, in one form or another, provide deserving workers with the resources they need to continue their work peacefully, sheltered from importunate needs.[53]

But this duty conferred no right of oversight or regulation,

> any more than subsidies to religion give the state the right to lay down the articles of faith. In one sense, the state can do even less with science than it can with religions; it can at least impose certain internal police regulations on religion, whereas it can do nothing, absolutely nothing, with science.[54]

Nowhere, in fact, was the support of scientific research regarded as an integral part of government responsibilities. In France the authorities were to continue living on the impetus of the reforms introduced by the revolution and consolidated by the Empire, without giving any large financial support to scientific research; Pasteur's working conditions in the rue d'Ulm are sufficient commentary on the limits of that support. The famous article written for the *Moniteur* and refused for publication, in which Pasteur spoke of laboratories as 'temples of the future, of riches and well-being', ended with an indictment of the shortcomings of the authorities:

> Who will believe me when I assert that there is not, in the national educa- tion budget, a single centime appropriated to the progress of the physical sciences by laboratory work; that it is only by means of a fiction and administrative tolerance that scientists, in their capacity as teachers, can draw some of the cost of their personal work from the public purse, at the expense of the appropriations intended for their teaching costs? [55]

For the rest, it was an international subscription and not the govern- ment which made it possible to create the Institut Pasteur after the success of vaccination against rabies.

Even in Germany, where the universities were increasing the number of well-equipped laboratories, the central government refrained from

53. *Ibid.*, 251–2. It is true that these principles are asserted after an intro- ductory paragraph which tells us a lot about the 'innocence' of the researcher in his relations with power: 'I should be departing from my plan if I were to hazard any ideas here as to practical applications. Furthermore, my complete ignorance of real life renders me quite incompetent to do so. Organisation de- mands experience and the balancing of principles against existing facts and can therefore in no way be the work of a young man.'

54. *Ibid.*, 253.

55. *Oeuvres de Pasteur* (ed. J. Vallery-Radot), vol. 7, Masson, Paris (1939), 203.

any direct aid to scientific research. When, after years of discussion, Helmholtz brought about the founding of the 'National' Institute for physical and technical research, the government gave its blessing, but nothing more; a private gift from the industrialist Siemens financed the initial work.[56] While the process of industrialisation was developing, and the progress of science from one great exhibition to another held out constant promise of more and more applications, the public authorities hung back from organising or directly aiding scientific research. Indirectly, it is true, they contributed to the progress of science by supporting, in Germany and later in Britain, the higher education reforms which were to confirm the professionalisation of scientists and to increase their numbers; but scientific research as such was not within their competence. In countries such as Britain and the United States where most of the universities were still private, intervention by the central government in the affairs of education and science was held to be a constitutional heresy. If it took a crisis—the 1914 war—for the heresy to gain countenance, government competence in this field remained very limited, and became even more limited immediately after the hostilities. In continental Europe, where the tradition of centralisation prevailed, the reluctance of governments may seem even more surprising.

There are at least two reasons for this reluctance, one connected with the place of science in the scale of social values and the other with the state of advancement of science itself. It is, moreover, reasonable to think that these two reasons are themselves to some extent interconnected; if scientific research suffered from the predominance of classical education, grounded on a general conception of culture which eliminated any utilitarian character, there was also the fact that science could not be counted upon to provide immediate applications. In fact research activities appeared to be an end in themselves in the context of a university culture, insulated by its institutions and context from community problems and mundane affairs; they were only on the fringe of university functions and remained there for such a long time in Europe that Jean Perrin could say, as late as 1933 that, 'the use of university grants for scientific research is an irregularity to which the authorities are prepared to turn a blind eye'.[57] The aspects of research

56. See the biography of Helmholtz by L. Koenigsberger, Brunswick (1902–3), cited by J. Schmandt, 'The Scientific Statesman' in *Les Etudes philosophiques*, P.U.F. (April–June 1966), 171.

57. See the biography of Jean Perrin by F. Lot, Seghers (1963), cited by Schmandt, *op. cit.*, 176.

capable of interesting the state, affecting its means of action or influencing the relative positions of one country and another, were regarded purely as a spontaneous, random and autonomous by-product of the sciences. If the state took over the functions of the ancient patrons, it was by subsidising scientific research as one cultural activity among others. And when it did so, always parsimoniously, it supported pure research rather than applied research, which was excluded from the conception of culture by its utilitarian bent.

In the United States, the diametrically opposed system of the time was based no less on the same dichotomy; because pure research was part of the cultured pattern, the government left it alone, applied research, especially in agriculture, being a matter for the Federal Government or the States according to the territory to be conquered and exploited.[58] As Toqueville pointed out,

> In America, the purely practical part of the sciences is admirably cultivated and great care is taken with the theoretical part immediately necessary to application; on this side the Americans display a spirit which is always clear, free, original and fertile, but there is practically no one in the United States who concerns himself with the essentially theoretical and abstract part of human knowledge.[59]

In the United States, the government still presumed that the union of applied research and free enterprise would be enough to guarantee progress; in Europe, governments imagined that progress must make do with a share of largesse assigned to pure research alone.

This cultural approach to science was precisely what confirmed the scientists in their one-sided conception of 'laissez-faire'. It was indeed the duty of the state to support science, but the state acquired no right in exchange, scientific activities being no less the realm of freedom than artistic activities. In the words of Renan himself, the priests of the new religion expected to be 'presented a living' just like their predecessors:

> The most natural way of granting patronage to science is thus by means of sinecures. Sinecures are essential in science; they are the most worthy and the most suitable way of giving the scientist an income, as well as having the advantage of grouping illustrious names of great ability around scientific establishments.[60]

58. See Don K. Price, *Government and Science* (1954; Oxford University Press edition, 1962), and A. Hunter Dupree, *Science in the Federal Government* (1957; Harper Torchbooks edition, New York, 1964).

59. A. de Tocqueville, *De la Démocratie en Amérique* (ed. Mayer), Gallimard, Paris (1951), vol. 2, 14.

60. *L'Avenir de la Science*, 254–5.

In the nineteenth century science demanded the blind support of the state because the state ought to feel its sole interest to be that science should flourish as a result; but, by the same token, the state had no reason to look for any advantages other than those it derived from subsidies granted to other cultural activities, such as the theatre or ballet. The doctrine of 'laissez-faire' held that even the slightest intervention was always for the worst—the least evil, otherwise, being that the state should grant its support blindfold.

The other reason for the state's meagre enthusiasm in supporting science is that there was still a long interval between scientific research and its applications, or more accurately that the link between science and technology was still loose, technology in fact making progress without relying directly on science. Not that science did not make enormous progress at that time; the nineteenth century was no less rich in essential discoveries than the twentieth, and these discoveries followed one another, contrary to what is often said, just as fast as today. If we take as a reference period merely the fifteen years from 1859 to 1873, it is clear that the scientific breakthroughs of the period lacked nothing in comparison with those of the last twenty years, either in importance or in the speed with which they followed one another, and this is even clearer if we take the years at the turn of the century as our standard of reference.[61] These scientific theories and discoveries represent a period of intense germination whose full fruits were not to become apparent until later—to such an extent that, in the light of their long-term implications for the technological developments we are witnessing today, it seems hard to find the contemporary equivalent of such a series of fundamental breakthroughs.

But the point to be made is that most of these breakthroughs failed to find short-term applications (apart from the work of Kekulé and, above all, Pasteur). For example, it took no fewer than forty years to progress from the discovery of electromagnetic induction to the indus-

61. The theory of evolution (Darwin, 1859), theory of sight and hearing (Helmholtz, 1860–2), spectroscopic analysis (Kirchhoff–Bunsen, 1859), theoretical structure of chemical compounds (Kekulé, 1858–66), microbial theory of disease (Pasteur–Kock, 1866–8), theory of heredity (Mendel, 1865), theory of electromagnetism (Maxwell, 1873), discovery of the electron (Thomson, 1897), discovery of radium (Curie, 1898), foundations of geometry (Hilbert, 1899), quantum theory (Planck, 1900), conditioned reflexes (Pavlov, 1904), theory of the subconscious (Freud, 1904), special relativity (Einstein, 1905). For this parallel with the pace of contemporary discoveries, see Hendrick W. Bode, 'Reflections on the Relation between Science and Technology', in *Basic Research and National Goals*, National Academy of Sciences, Washington (1965), 48f.

trial perfecting of the dynamo, and some thirty years to go from the discovery of the formula of benzene to the large-scale production of industrial dyestuffs. Furthermore, a great many of the technical achievements of the nineteenth century were arrived at independently of the scientific knowledge on which they were based. The classic, though by no means the only, example is the steam engine, of which Watt had the first idea as early as 1765, which was perfected thanks to the use of high-pressure steam in 1802, and led from 1825 onwards to the use of locomotives; whereas it was the middle of the century before the laws of thermodynamics, foreshadowed by Carnot in 1824, were completed and generalised by Clausius and Kelvin. The Industrial Revolution was bound up more closely with the progress of technology than with the progress of science, and science itself up to the end of the nineteenth century was to depend more on industry than industry on science.[62]

In one sense, Auguste Comte was right: the temporal lords, bankers and industrialists, played into the hands of the scientists. As proved by the example of Germany, where the first big applied research laboratories developed, industry rather than the political authorities assumed responsibility for scientific research in the nineteenth century. But the scientists did not, for all that, become the spiritual leaders. Comte believed that industrial societies would make war an anachronism: since there would no longer be a military class, there would no longer be any reason to fight; the age of the industrialists would put an end to the age of armed conflict. He did not foresee that scientists, in association with industrialists, would take a greater part in wars than ever, and that many of them would equal the soldiers (and the politicians) in their failure to display 'the characteristic repugnance of modern societies for the life of war'.[63]

62. See Charles Singer *et al.*, *A History of Technology*, Oxford (1958), vol. 4 (1750–1850) and vol. 5 (1850–1900), and, especially, J. D. Bernal, *Science and Industry in the Nineteenth Century*, Routledge and Kegan Paul, London (1953).

63. *Cours de Philosophie positive* (ed. Schleicher Frères), Paris (1907–8), vol. 4, 375; on this point, see R. Aron, *La Société industrielle et la Guerre*, Plon, Paris (1952), 1–82.

CHAPTER 2

The Turning Point

The first world war brought science and politics into a closer relationship, without greatly changing the nature of the relationship from that brought about by the great armed conflicts of the nineteenth century such as the American Civil War or the Franco-Prussian War. A few scientists were engaged as such in advising governments about military programmes based on scientific and technical achievements. Military research was confined mainly to adapting available knowledge and techniques to the needs of the armies. There was nothing really new, in spite of the novelty of some of the arms and techniques used: tanks, submarines, balloons, dirigibles, airplanes or gas, with the techniques of detection and protection they implied, transferred to the battlefields instruments already used, in one form or another, in the civil sector. The biggest difference between the 1914–18 war and earlier wars was not so much the use of new technologies—most wars have been 'midwives' to new techniques—as the recourse to mass production (and mass mobilisation); from this point of view, the first world war effectively consolidated the era of industrial societies in Comte's sense, that is to say, the application of science to the organisation of labour.

There would be no point in mentioning this experience of the first world conflict, if it did not demonstrate the fact that state commitment to scientific research was still limited. It rarely went beyond applied research and, above all, industrial research. In pure research, the record of all institutions created during the war seems to be very modest; according to some people, it is even a negative quantity, the majority of scientists engaged on military research devoting themselves purely to applied research: Frank B. Jewett, himself prominent in war research in the United States, claimed that

> in setting up the machinery to accomplish these [recent scientific wartime] achievements we at the same time set up the machinery for the destruction of advances beyond a certain point.[1]

1. Quoted by A. H. Dupree, *Science in the Federal Government*, Harper Torchbooks, New York (1964), 324.

After the war, the relations between science and politics reverted in most countries—at any rate until the 1930s—to what they had been in the second half of the nineteenth century, and can be summed up as 'good neighbour' relations in which the state supported research as a superfluity and science was in no position to insist on the essential.

If all these institutions, before and after the war, played a very limited role in coordinating research, it was because neither the government machinery nor the scientific organisations were ready to assume such responsibilities. The scientists refrained from taking part in political decisions and the politicians had no idea that science could affect politics. Politicians and military men looked on scientists as strangers to the world of action, politics and war; the scientists mistrusted the administration and above all refused to admit into their laboratories any rumblings of political battles, even where they were not quite simply deaf to them. Many scientists had the same reactions as J. Robert Oppenheimer during his formative years at Berkeley:

> I was not interested in and did not read about economics or politics. I was almost wholly divorced from the contemporary scene in this country. I never read a newspaper or a current magazine like *Time* or *Harper's*. I had no radio, no telephone; I learned of the stock market crash in the fall of 1929 only long after the event; the first time I ever voted was in the Presidential election of 1936. . . . I was interested in man and his experience; I was deeply interested in my science; but I had no understanding of the relations of man to his society.[2]

Science militant

In fact, the fears felt by certain scientists were not so much that their autonomy might be limited by undue demands from the state as that science might become politicised owing to their own abdication as researchers. The threat was not that science would be treated in the political sphere as a means rather than an end, but that the whole of politics, ends and means, would be shifted by the scientists themselves into the scientific sphere, or, in other words, that science would be liable, in Max Weber's words, to self-sacrifice for the benefit of believers, fanatics and prophets.[3]

Max Weber's two famous addresses in which science and politics were defined as antagonists, the theatre of a fight to the death between opposing idols, each possessing a different vocation and a different ethic,

2. *In the matter of J. R. Oppenheimer*, transcript of the hearing, U.S.G.P.O., Washington (1954), 8.
3. Max Weber, *Essays in Sociology* (trans. H. H. Gerth and C. W. Mills), Oxford University Press (1970), 125.

were delivered in the winter of 1918. On the morrow of defeat Germany was torn by partisan conflicts and Munich was the scene of bloodshed; here Weber sought to make the voice of reason heard, a voice to which men were already beginning to turn a deaf ear. These addresses are powerful warnings against the risk incurred by scientists and politicians of 'letting themselves in for the diabolic forces lurking in all violence'.[4] At least the politician, as a professional, must answer to the gods. Of course, 'an ethic of ultimate ends and an ethic of responsibility are not absolute contrasts, but rather supplements';[5] but even so the politician must be a hero to reconcile them and live with them in reconciliation without yielding to the temptations of fanaticism or violence. Between the citizen of Florence who prefers, according to Machiavelli, the salvation of the city to the salvation of his soul, and Luther voicing his unconditional rejection of the pitfalls of history in his declaration to the Diet of Worms—'Here stand I: I can do no other: God help me'—Max Weber recognises that politics, the scene of action and of choices, demands its own sacrifices, and that a man must be of heroic stature to carry sacrifice to its culmination. But if the scientist, as scientist, must aspire to heroism, it is the heroism of truth and not of history.

The intense fever of nationalism which swept Europe before and after the first world war did not spare the world of science; the war ended, the fever was counteracted, replaced or prolonged by the hope of the socialist revolution:

> Not summer's bloom lies ahead of us, but rather a polar night of icy darkness and hardness, no matter which group may triumph externally now.[6]

We shall see below how the second world war committed researchers to politics in a way which Weber did not foresee, and how the ethic of responsibility laid siege to the ethic of conviction instead of the ethic of conviction balancing the ethic of responsibility, as he hoped would happen in the political sphere. We must be content for the moment with emphasising that between the two world wars, the scientists—at any rate in the democracies—were in danger less of having to bend their research to an authoritarian orientation than of yielding themselves to a partisan faith.

In this respect, there is no fundamental difference between the natural sciences and the human sciences: the line of resistance to fanaticism or violence depends not on the nature of the discipline, but on the man. The only difference is that specialists in the human sciences are more

4. *Ibid.*, 125. 5. *Ibid.*, 127. 6. *Ibid.*, 128.

'sensitised' to events and to politics; history is nearer to them, because they must inevitably take account of it in their work. For the natural sciences, history is another world; the 'diabolic forces' are still strangers to the laboratories, because natural scientists, from innocence, lack of interest or absorption in quite different concerns, are strangers to them.

But the 1930s were to involve many scientists, hitherto indifferent to politics, in an awareness and sometimes in a commitment or allegiance. On the side of the democracies, it was the economic crisis which first revealed the cost of 'laissez-faire'; the idea that progress could result only from the free union of science and industry collapsed under the blows of the depression. For some people the revelation was their discovery of a responsibility inherent in their professional activities in view of the social implications of those activities, the control of which they felt they could or should share with politicians outside the laboratory and the university. And yet the recognition of the role which science could play in social and economic development did not go as far as endowing the state with means of orienting the research effort or organising it more coherently.

In the United States, Roosevelt called in the counsels of the Science Advisory Board in 1933, in order to associate researchers with the New Deal. Once again, the interest shown by the government in scientific activities was not accompanied by the recognition of any responsibilities: the Science Advisory Board was financed not by the national budget but by private foundations, and its work was limited to drawing up reports and recommendations which had no sequel. The days of 'laissez-faire' were ended only on paper: if science was defined as 'one of the greatest resources of the nation', in the words of the President of the United States himself in his review of the research effort,[7] the committees which pleaded its cause at government level were still without the administrative status or the support from scientific organisations which would enable them to influence researchers and political decisions. The outstanding achievement of the New Deal in this sphere was to attract to Washington scientists such as Karl T. Compton, James B. Conant and Vannevar Bush; having served their time in the antechambers of political power, they were to have access to the President from the outbreak of the second world war.[8]

7. Letter from Franklin D. Roosevelt to F. A. Delano, July 1937, in *Research—A National Resource*, Washington (1938), vol. 1, 2.
8. See Dupree, *op. cit.*, 368, and Lewis E. Auerbach, 'Scientists in the New Deal: a pre-war episode in the relations between science and government in the USA', *Minerva*, 3 (summer 1965).

It was in Europe, under the totalitarian régimes, that politics laid its hands on science from the outset, and on two planes, ideology and the realities of action. In Soviet Russia, just as in Nazi Germany, political power subordinated science to the orthodoxy of which it boasted; science became politicised through the very fact that politics decided the affairs of science. The relativity theory, like psychoanalysis, was dismissed in the USSR in the name of Marxist science, and in Germany in the name of Aryan culture. But the transition from ideology to action had very different consequences in Germany and Russia since the two ideologies entertain diametrically opposed concepts of science. Communism holds that science and politics are joined together because they both obey, in theory, the same rationality; Nazism, in contrast, tends to separate them on the very grounds of the irrationality which it professes.[9] In a certain sense, while it is true that the construction of Communism lay through science, the destiny of Nazism lay through the destruction of science: the price paid by German science (the emigration of researchers, the cumulative lag in information and training, the disruption of university structures) was heavy indeed.[10]

At this point a fundamental difference appears to have emerged between the natural sciences and the human sciences; it was no longer a question of researchers as men or citizens, more or less sensitised to world affairs, but of the whole nature of scientific disciplines, more or less vulnerable to the pressures of ideology and political power. The sciences of man—especially the social sciences—steeped in history and nourished by it, much as they may disavow it—cannot completely free themselves from the weight of the philosophies which inspire events. Truth must always wage 'a more doubtful combat' on this field than on the field of the natural sciences. This does not mean that the issue of the combat is more doubtful; there are also, as Raymond Aron has recalled, constitutional rules of the 'community of social sciences' which

9. Among many others along the same lines, this statement in 1936 by Bernhard Rust, Reichsminister of Education, at the 550th anniversary of the University of Heidelberg: 'National Socialism is justly described as unfriendly to Science if its valuer assumes that independence of presuppositions and freedom from bias are the essential characteristics of scientific inquiry. But this we emphatically deny.' Cited by B. Barber, *Science and the Social Order*, Collier Books, New York (1962), 112.

10. On the consequences of the Nazi adventure in the scientific field, see Leslie E. Simon, *German Research in World War II*, John Wiley and Sons, New York (1947); Samuel A. Goudsmit, *Alsos*, Henry Schuman, New York (1947); and Joseph Haberer, *Politics and the Community of Science*, Van Nostrand Reinhold (1971).

restrain science from turning into mythology or propaganda.[11] But obeying these rules is not enough to ensure that there is a consensus about the truth. The natural sciences do not run the same risks; they can be used as propaganda but they cannot be referred to as mythology. If physicists, chemists or biologists can be compelled to bow to orthodoxy, orthodoxy cannot make their theory right if it is wrong or make an experiment prove what it cannot prove. 'Whereas in a Congress of Philosophy', said Bachelard, very wisely, 'philosophers exchange *arguments*, in a Congress of Physics, experimenters and theorists exchange facts.' [12] Nothing, for example, will stop the theory of pauperisation from having its partisans (and its arguments) a century from now, but orthodoxy did not prevent Heisenberg in Germany or Kapitza in Russia from using Einstein's calculations—and Lysenko's theories did not survive the death of Stalin.

Now, in spite of this difference, the tumult of the congress of philosophers was to invade the congress of physicists. The 'mounting peril', the series of crises which paved the way for the second world war (the Spanish civil war, Munich, the persecution of the Jews), threw some scientists into the political battle; intellectuals among their fellows, but representing science, they added the weight of their reputation, or simply of their calling, at the foot of petitions, in national and international gatherings, in the service of parties. The attractions of Communism, the shadow of Fascism, led to the ideological mobilisation of science and split the scientific community. In the polemic between left-wing and right-wing intellectuals, scientists discovered in their turn that, in espousing a cause, it is not easy to distinguish the expert from the citizens, and that it is still less easy to distinguish the support given from its political exploitation.

Communism and science

The arguments set off by Nazism on Aryan culture scarcely called for refutation, since they were essentially based on non-science. The debate that was opened by the very existence of a state whose leaders set up their philosophy as a science had a very different significance. The fact was that the revolution triumphant in Russia consolidated the closest relation that had ever existed between science and politics. The transition from ideology to action here provided a model of organisation; not

11. Introduction to Max Weber, *Le Savant et le Politique*, Plon, Paris (1959), 22–3.
12. G. Bachelard, *Le Rationalisme appliqué*, P.U.F., Paris (1949), 1.

only was science recognised as a national capital asset, but it was also proclaimed a public service and integrated in the forces of production. Scientists influenced by Marxism could point both to theory and to social practice, which made the Soviet state the unstinting patron of science and science the whole-hearted servant of the state, as something quite without parallel in the liberal régimes.[13]

Non-Marxist researchers were not slow to discover that the growing dependence of the economic and social system on scientific research was no less linked with state intervention in the liberal régimes. But no one, on the eve of the second world war, thought of looking for characteristics common to the democracies in which the doctrine of 'laissez-faire' prevailed and to the Soviet régime which asserted the need to integrate science into the whole production process. The Communist ideology determines in theory a radically different form of society in which all the means of production are in the hands of the state, in which the state proclaims its duty to support science and at the same time its right to orient science and use it in the context of overall planning. In this light, the interests of science are the same as those of the political authorities, since the objectives of the one are at the same time the objectives of the other. The centralisation of decision-making enables the state to draw up an overall plan, to set the pace of development, to speed up the progress of sectors deemed essential or more valuable for the power of the community. The organisation of scientific research cannot escape the uniform criteria of state control and planning, any more than other sectors of production; in a sense, it should escape them less than other sectors since the path of science is defined as one of the most direct means of access to Communism.

But reality was far from matching this theoretical scheme, and still further from the idyllic vision formed, under liberal régimes, by left-wing intellectual scientists such as J. D. Bernal:

> The cornerstone of the Marxist State is the utilisation of human knowledge, science and technology for human welfare.[14]

There can be no doubt that scientific activities enjoyed a status and support in the USSR which had no parallel in any other country before the second world war; the scientific system was inseparable from the political system, of which it was both the means and the end. And yet the integration of scientific research into the production process did not

13. Especially J. D. Bernal, *The Social Function of Science*, Routledge and Kegan Paul, London (1939; re-issued in 1967 by the MIT Press).
14. *Ibid.*, 22.

seem to be so easy or so free from conflict as the official literature and the party pundits proclaimed. The repeated reshuffling of the organisation of scientific research in the USSR after the reconstruction period was enough to show that, even if the cause of the state and the cause of science were wedded, the honeymoon was not necessarily idyllic.[15]

Leaving aside the human sciences which, by definition, are easier to bring into line, the planning of the natural sciences led to difficulties of all kinds, illustrated in particular by the constant efforts of the Academies to keep research unsullied by short-term planning objectives and the reluctance of the economic system to accept technical innovations. The situation in Russia at this time was very similar to that in France under the revolution: state intervention concentrated mainly on applied research and brought about major changes in the higher education system, the results of which became apparent only later. The planning of science, moreover, did not escape either the pressure of economic events or the defects of economic management so highly centralised that it could not take into account the initiatives of the research units situated at the base of the pyramid of decision. These units, under cover of the short-term programmes included in the plan, did not hesitate to continue and even to launch other programmes not aimed at any immediate application or not included in the plan.[16]

The power of decision in scientific matters was, of course, unquestionably more concentrated in the USSR than in any other country in the world, but state control did not stop some fundamental and even applied research institutes from approaching questions from the individual angle rather than in the bureaucratic framework of a specific assignment. Kapitza has told how he never researched except on subjects of his own choosing.[17] Similarly, the leading design offices in the aircraft industry were built up around gifted engineers whose struggles with party bureaucracy were no less arduous than those of Western innovators with the management of private companies; Yakovlev, the progenitor of the series of Yak fighters, had as much difficulty in persuading his works manager to allocate him premises for his design

15. See *Science Policy in the USSR*, OECD, Paris (1969), and especially the analysis of the Academies by H. Wienert and of science in the economy by R. W. Davies and others.

16. See J. M. Colette, *Recherche-Développement et Progrès économique en URSS*, Cahiers de l'I.S.E.A., Paris (1962).

17. *Peter Kapitza on Life and Science*, The Macmillan Company, New York (1968).

office as Frank Whittle had in inducing the company which employed him to give him money to continue his work.[18] Even more surprisingly, the research and development offices were organised and financed independently of industry or, in other words, of production.

The ideology which reduces scientific activities to productive activities among others assumes that there is no policy or specific problem concerning support for research or the exploitation of research results throughout the economic system. But, in reality, research and development activities are more radically separated from production than in the capitalist economies. The state in which science is most officially recognised as a national asset and most closely subordinated to centralised decisions still has no science policy as such. It is striking to find that it was from contact with the work developed in the United States since the second world war on the relations between science, public affairs and the economy, that the Soviet Union discovered, no less belatedly than Western Europe, the original character of the research system in the production process and the specific nature of science policy in its range of bodies and decision-making.[19]

If practice does not match theory, it is also because the underlying ideology can find no precise rules in Marxist doctrine for the conduct of scientific activities as such. In asserting that the appropriation by the state of the means of production changes the working conditions of the researcher whose motives and objectives are different because he is contributing to the construction of Communism rather than to a system based on profit, or that state planning of science makes it easier to apply science to the problems involved in achieving the objectives of the community, the apologetic literature—that of left-wing scientists in non-communist countries as well as the official literature—is pursuing a sort of tautological reasoning in whose terms the practice of science is subordinate to the ends of society and the ends of society are subordinate to the practice of science.[20]

18. R. W. Davies, 'Science et Economie en URSS', *Atomes*, Paris, no. 247 (October 1967).

19. See G. M. Dobrov, *Nauka o Nauka (The Science of Science: Introduction to the General Study of Science)*, Kiev, Naukova Dumka (1966). This discovery based on American references is even more manifest in the report of the research team directed by Radovan Richta, *Civilisation at the Crossroads: the social and human implications of the scientific and technical revolution*, Prague, Academy of Sciences (1967).

20. For example: 'The Soviet scientist clearly realises to what end he works. He is entirely free of doubts and disillusionments caused by contradictions between his intellectual aspirations and the ideals of humanism. The Soviet

Now, this literature always refers back to the same passages in Marx or Engels, who, in the first place, do not deal directly with the question of the relations between science and the state and, secondly, define science as an activity solely dependent on industry and technology. The politicisation of science—the intrusion of politics into science, not only as the scene of decision and action by the state, but also as the theatre of rivalries between different philosophies—was no doubt inherent in Marxist thinking from the outset, since the class struggle is no less present in this field than in that of production in general, and yet nothing is more certain than that Marx's writings tended to create a division of this kind in the heart of science itself.

The determinism which Auguste Comte saw at work in the course of history, forcing mankind to attain the positive stage after the theological and metaphysical stages, is equally irresistible for Marx, but it refers back to a quite different conception of the driving forces of history and the ways in which history fulfils itself. Whereas for Comte the destiny of societies is determined by the progress of the human mind, for Marx it is the forces and relations of production which determine the historical process.

> The mode of production in material life determines the general character of the social, political and spiritual process of life. It is not the consciousness of men that determines their existence, but on the contrary, their social existence determines their consciousness.[21]

From this point of view, in which the infrastructure of societies, that is to say, their economic basis, conditions men's social patterns and ways of thought, scientific activities, the product or reflection of a given state of the production system, are at the same time a constituent of the forces of production.

Since the essence of these forces lies in the given capacity of a society to produce, a capacity which depends on the technical apparatus and the organisation of labour, but also on scientific knowledge, there is no distinction between the theory of science and its social practice, and it is therefore part of the infrastructure; but scientific activities are at the

---

scientist is not tortured by the idea that his discoveries might be used for the enrichment of others, for the exploitation of people and the annihilation of people.' N. A. Figurovsky, 'The Interaction between Scientific Research and Technical Invention in the History of Russia', in *Scientific Change* (ed. A. C. Crombie), Heinemann, London (1963), 721, or J. D. Bernal, *The Social Function of Science*, 224, and the same author's *Marx and Science*, International Publishers, New York (1952), 42 (in particular).

21. Marx, *Critique of Political Economy* (1859).

same time part of the superstructure since, at any given moment in history, they form part of human consciousness, a factor in the development of intellectual life. Without stopping at basic Marxism, which asserts that the superstructure is automatically determined by the economic situation, the fact remains that the substratum of economic conditions provides the master key to the understanding of the history of science, even in its most theoretical aspects. There is a choice: either the history of science can be written only at the cost of generalisations which are as vague as they are dogmatic, since the genealogy of concepts and theories always depends, in the last analysis, on conditioning by the economic system; or science must be defined in such a way as to place it from the outset among the ranks of determinant forces rather than of ideologies determined by those forces. In the first case, the history of the sciences vanishes; in the second case the sciences do not exist except in so far as they are treated as merely one aspect among others of the forces of production.

This is yet another example of the ambiguities of the Marxist theory of knowledge, and dialectics do not suffice to eliminate it by the mere fact of establishing a constantly reciprocal relation between the movement of ideas and social facts. One must accept that there is an inherent contradiction in scientific activities, similar to that which is found in art, about which

> it is well known that certain ages in which art flourishes bear no relation whatever to the general evolution of society or consequently to the development of the material basis which is, as it were, the framework of its organisation.[22]

Otherwise the sciences can have no independent history of their own and are governed by the same process of determination as industry and technology of which they are the direct offspring:

> Feuerbach speaks in particular of intuition in the natural sciences. He talks of mysteries perceptible only to the eyes of the physicist or the chemist; but where would the natural sciences be without industry and trade? Even these 'pure' natural sciences in fact get their aims and their materials through trade and industry, through the sensible activity of men.[23]

Marx saw very clearly what industry owed to the development of technology, particularly when he emphasised, in connection with machine-tools, that:

22. Marx, *General Introduction to the Critique of Political Economy*.
23. Marx, *German Ideology*.

The difference strikes one at once, even in those cases where man himself continues to be the prime mover.[24]

More than that, he foresaw, in a technological landscape which was only just beginning to take shape—that of steel, steam engines and chemical fertilisers, very recently introduced by Liebig's work—the productive role of knowledge as a substitute for labour itself. This anticipation is found in a striking passage in the *Grundrisse*, in which the industrial process, enrolling all the sciences in the service of capital, transforms the intervention of man in inorganic nature into a 'spontaneous act':

> Invention then becomes a specialty and the application of science to immediate production becomes a determinant concern and postulate for the inventor.[25]

The development of the sciences determines a new form of industrial development in which

> the worker no longer intervenes as a link between the natural object and himself . . . [but] takes his place alongside the production process instead of being its principal agent.[26]

All knowledge becomes immediate productive power—'the materialised power of knowledge'[27]—and existence can hope to free itself from the servitudes of labour; a civilisation of leisure is made possible by this transformation of the industrial process in the face of which man is called upon to act as 'supervisor and regulator'.[28]

But this vision of the future did not go so far as to invert the relation between science and industry and make industry the offspring of science. If, in Marx's view, 'it was through industry that the natural sciences came into human life in a practical way',[29] it would henceforth be through the natural sciences that industry entered the destiny of mankind. Pure science played a minor role in the Industrial Revolution, and while it is true that Marx visualised the importance of the applied sciences, especially for agriculture as a result of Liebig's research, the function he assigned to them was related more closely to the industrial process than to pure science. Only in the nineteenth century did the progress of certain scientific theories begin to affect the development of industry, but the progress of machinery was only the start of the 'science-based industries'. The relation was finally to become inverted to

24. Marx, *Capital* (English translation by Moore and Aveling), Lawrence and Wishart, London (1970), vol. 1, chapter 15, 374.
25. Marx, *Principles of a Critique of Political Economy*.
26. *Ibid*.   27. *Ibid*.   28. *Ibid*.
29. Marx, *Outline of a Critique of Political Economy*.

such an extent that nowadays it is rather the progress of industry which depends on scientific theories.

In fact, Marxist thinking endows science with an ambiguous status precisely in so far as it reduces it to the role of a servant of industry. Even if, as the 1844 manuscripts assert, sensibility is the sole genesis of knowledge, this alone is not enough to account for the whole history of knowledge; the status of science hesitates between the sensible determinations of the infrastructure and the subjective conditioning of the superstructure. Sometimes the natural sciences are associated with the economic process of industry and regarded as natural forces whose determination becomes apparent only from the moment they are exploited:

> A water-wheel is necessary to exploit the force of water, and a steam engine to exploit the elasticity of steam. Once discovered, the law of the deviation of the magnetic needle in the field of an electric current, or the law of the magnetisation of iron, around which an electric current circulates, costs never a penny.[30]

In this sense science is, in a way, 'outside the economic circuit', defined as the system of available knowledge which costs nothing, such as 'physical forces, like steam, water, etc., when appropriated to productive processes'.[31] But regarded as part of the production system, the natural sciences are equally tainted with ambiguity, since they intervene simultaneously or in turn as a determinant system or as a conditioned system, as the objective purpose of labour and the subjective history of consciousness.

The assertion of the close link between theory and practice, or even the assimilation of theory with practice, does not prevent science from being, in a sense, part of the superstructure. The debate on what belongs to the forces of production and what belongs to the realm of ideas can go on for ever, like the debate about how the realm of ideas is conditioned by the forces of production. Marxist thought has left an ambiguous heritage, which opened the door to furious battles within the Soviet intelligentsia, as exemplified by the debate which raged in the inter-war years between the 'mechanists' and the 'Deborinists'. The Deborinists thought that science was closer to the infrastructure, thus reflecting a class phenomenon, like philosophy or art, and dividing scientists according to their allegiance to Marxism or their bourgeois associations. For the Deborinists it was closer to the superstructure, introducing among scientists, whatever their class or their ideology, the

30. Marx, *Capital, op. cit.*, 386.   31. *Ibid.*

bonds of a factual community, since dialectical materialism was defined in their eyes as inherent in the method and spirit of the natural sciences, and in the last analysis all scientists, from the mere fact that they obeyed this method and spirit, were materialists, albeit unconsciously.[32]

Against this background of ambiguity, science regarded as a system of knowledge is no less related than science regarded as a social institution to the conflicts between the forces of production and the relations of production, that is to say, the relations of property and the distribution of income among the individuals or groups of the community.

> The point of view of the old materialism is the bourgeois society, the point of view of the new is human society or humanity.[33]

Marx still speaks only of a point of view, where Lenin was to proclaim the official doctrine of the party. The link between materialism and Communism is the link which is to bind the destiny of science to the destiny of the proletariat, but Marx nowhere says whether this link involves the subordination of science to the state. In theory, since science does not escape from class antagonisms, the assumption of power by the proletariat and the end of antagonisms should mean the emancipation of science from political power as such. But the golden age of Communism fulfilled, when 'the public power will lose its political character' as 'the organised power of one class for oppressing another',[34] still has to prove itself. In the meantime, the last word on the relationships between science and the state lies with the party, whose interpretation of dogma varies according to circumstances, according to scientific disciplines, and even according to individuals, in the light of the services rendered by scientists to the community, sometimes even in spite of the liberties they take with orthodoxy. The ideology which contributed most to the politicisation of science after the first world war begged a question which Marx had not put, and divided the sciences against

32. See David Joravsky, *Soviet Marxism and Natural Science, 1917–1932*, Columbia University Press (1961), and Loren A. Graham, *The Soviet Academy of Sciences and the Communist Party*, Princeton University Press (1967). The group of 'Deborinists' was named after Abraham M. Deborin, who led the fray as the most influential academic philosopher against the 'revisionists' (Stepanov, Bogdanov, Axelrod, Varjas and others) who, without forming a very coherent 'faction', relied on the spirit rather than the letter of Marxism in interpreting dialectical materialism. The controversy was all the more important since it involved scientific circles, in physics on the relativity theory and in biology on Morgan's theories (Joravsky, *op. cit.*, chapters 18 and 19).
33. Marx, *Theses on Feuerbach, German Ideology*.
34. Marx and Engels, *Communist Manifesto*.

themselves in the light of their subservience to a dogma unknown to Marx.

In spite of its ideological ambiguities and its political constraints, the Soviet experience of the relations between science and the state was to be the model for left-wing scientists in the non-Communist countries. Under the mounting threat of Nazi Germany and Fascist Italy, the campaigns in favour of support for scientific research, of its recognition if not as a public service then at least as an investment of national importance, were in harmony with the defence of the democracies. These campaigns pointed to the Soviet state as an example of the convergent interests of science and politics, of what politics can contribute in the way of support and what science can contribute in the way of means. The short-term interests converged because of the threat of war, but the long-term interests converged as well because of the social role of science as an arsenal of resources on which the state had only to draw in order to solve its social and economic problems. The socialist vision of the social role of science joined hands with that of the century of the Enlightenment, in which knowledge and power should act in concert for the happiness of mankind.

Among the democracies, France was the only country in which this convergence was recognised and institutionalised before the second world war. The Popular Front celebrated the betrothal of science and politics, announced by the attendance of Jean Perrin at the mammoth demonstration of 14 July 1935:

> Citizens and comrades—Communists, Socialists, Radicals—and all you men of good will united above party by the same determination to defend your freedom . . . I come to greet you and to say 'adsum' since I am one of those to whom you have done signal honour by charging us to express your anguish, your enthusiasm and your faith in that better society which is made possible by science. For it is indeed liberative science which will create ever more wealth and beauty which must be extended to all men.[35]

On the strength of his old friendship with Léon Blum, Jean Perrin asked him, when he formed his Popular Front government, to affirm in his Ministerial statement

> the paramount importance of research and development for the progress and greatness of the nation. He answered me [recalls Jean Perrin] that he would do even better by making Irène Joliot-Curie an Under-Secretary, thus succeeding for the first time, by a choice to which there could be no

35. Quoted by Albert Ranc, *Jean Perrin, un grand savant au service du socialisme*, Editions de la Liberté, Paris (1945), 50–1.

possible objection in principle, in ensuring the representation in the government both of Frenchwomen and of scientific research.[36]

In fact, Irène Joliot-Curie, who accepted the post with reluctance, soon discovered that she was not cut out to be a minister, and Jean Perrin, another Nobel prizewinner, was invited to take her place. Thanks to his efforts as spokesman for science within the Cabinet, the Council for Scientific Research, created as early as 1933 but with only scanty resources from the Caisse des Sciences, and limited to expressing its wishes for the attention of the Minister of Education, was able to act with more powerful means.

This commitment of scientists to political decision-making resulted in the creation of various institutions, culminating in the *Centre national de la Recherche scientifique* in 1939. For the first time in a non-Communist country, the government assumed responsibility for fundamental research as such; the days of 'laissez-faire' were over and pure science had become an affair of state. Thus, the politicisation of science opened the way to science policies; the intrusion of politics into science as the theatre of rivalries between different schools of thought defined the new relations between scientific research and the state, under which ideological commitment legitimised both the intervention of the public authorities and the mobilisation of researchers. But this time might not have come, or at any rate not so soon, without the intrusion of science into politics as a decisive factor in the balance of power between nations; it took the second world war, the massive recourse to scientific research and, above all, the perfection of nuclear weapons, before politicised science extended itself and fulfilled itself in science policies.

The mobilisation of science

> Some recent work by E. Fermi and L. Szilard, which has been communicated to me in manuscript, leads me to expect that the element uranium may be turned into a new and important source of energy in the immediate future. Certain aspects of the situation which has arisen seem to call for watchfulness and, if necessary, quick action on the part of the Administration. I believe therefore that it is my duty to bring to your attention the following facts and recommendations. . . .[37]

By this appeal, addressed to President Roosevelt on 2 August 1939, Albert Einstein sought to bring home to the American government the military issues at stake in the results achieved during the previous four months by Fermi and Szilard in the United States and Joliot-Curie in

36. *Ibid.*, 39.
37. In *The Atomic Age* (ed. Grodzins and Rabinowitch), Simon and Schuster, New York (1965), 11.

France. It can be regarded, if not as the beginning of the 'atomic age', at least as the starting point of the new association between knowledge and power. It was certainly a 'historic' approach, which was to precipitate a no less radical change in the relations between peace and war than in the relations between science and politics, but there is nothing epic about its background. Einstein, who at the beginning of 1939 was still saying that he did not believe in 'the early release of atomic energy', merely acted as a post office.[38] The letter he signed was drafted by Szilard himself and was handed to Roosevelt only in October 1939 by Alexander Sachs, an economist and businessman and a personal friend of the President.

From the outset, the new relation between scientists and power, initiated by atomic research, was stricken with ambivalence. In the first place, by an irony of fate it was the most notoriously pacifist of all scientists who gave the signal for the construction of the most destructive weapon ever designed. There can be no doubt that, in agreeing to recommend to the President of the United States that the question of uranium should be actively studied, Einstein had no thought that the atom bomb, if by chance it could be realised, could ever be used except as a safeguard against the surprise of a German nuclear weapon. The group of physicists who feared that Germany would forestall the democracies in perfecting the atom bomb needed a spokesman with enough prestige to gain access to the summit of the American executive. It was later to be learned that the German scientists never really tried to make an atom bomb, either because they themselves did not think it possible, or because they went off on another line designed to equip Germany with nuclear reactors instead of atomic weapons, or finally because the Nazi leaders saw no value in research in this field.[39] In the eyes of those who perfected it, the atom bomb was to be a deterrent only and this

38. Many of Einstein's own writings (e.g. *Conceptions scientifiques, morales et sociales*, Flammarion, Paris (1952), 222) as well as *Le Drame d'Albert Einstein* by Antonina Vallentin, Plon, Paris (1954), confirm that he agreed to lend his name, but it is not generally known that he was extremely reluctant. Oppenheimer himself told me that Einstein thought an approach of this kind to Roosevelt was untimely, both because he doubted the possibility of making the atom bomb and because he foresaw the political implications if the research were to succeed.

39. See Robert Jungk, *Brighter than a Thousand Suns* (English translation by James Cleugh), Gollancz and Hart Davies, London (1958; reprinted in Pelican Books, 1964), Samuel A. Goudsmit, *Alsos*, Henry Schumann, New York (1947), and above all R. G. Hewlett and O. E. Anderson, Jr., *The New World: A History of the United States Atomic Energy Commission*, the Pennsylvania State University Press (1962).

deterrent was to be used only against Germany. But it was over Japan that the first nuclear bombs were dropped, and the same men who originated Einstein's approach to Roosevelt tried in vain to change the decision to make use of their work.[40]

There is no better proof of the ambivalence of this relation than the conclusions of the scientific committee responsible for making recommendations to the Secretary of War, Henry L. Stimson, for the choice of target. From a purely technical demonstration in a desert to the bombing of purely military sites, the experts could not reach unanimous agreement and passed the buck to the political authorities, on the very grounds of that scientific competence which some of them had invoked to intervene with the authorities six years before.

> We can propose no technical demonstration likely to bring an end to the war: we can see no acceptable alternative to direct military use. With regard to these general aspects of the use of atomic energy, it is clear that we, as scientific men, have no proprietary rights. *It is true that we are among the few citizens who have had occasion to give thoughtful consideration to these problems during the past few years. We have, however, no claim to special competence in solving the political, social and military problems which are presented by the advent of atomic power.*[41]

Trapped by the forces they had unleashed through their association with political power, the scientists could no longer escape from the ambivalence of a situation in which they appeared at one moment as the champions of an expertise which must be distinguished from the uncertainties of political decision and at another moment as the holders of political competence based on their privileged access to scientific questions.

Much has been written, and in great detail, about the history of the first nuclear bombs and the way in which they led researchers—first the 'atomic scientists', who had been directly associated with their perfection in the United States, followed by others in other disciplines and other countries—to intervene *ex officio* in the political field, individually and collectively. Our purpose here is not to recount this history, but merely to emphasise the significance of the resultant change in the relations between science and power. Among the scientists, the feeling of special responsibility arising out of the nature and consequences of

40. See James Franck and others, *A report to the Secretary of War* (better known as 'the Franck Report') (June 1945), and Leo Szilard, 'A Petition to the President of the United States' (17 July 1945), in *The Atomic Age*, 19–29.

41. Quoted by Henry L. Stimson, 'The Decision to Use the Bomb' (February 1947), in *The Atomic Age*, 35–6 (my italics).

their work was no doubt the decisive motive for political commitment;[42] science ceased to be neutral—if ever it had been—and scientists could no longer dissociate themselves from the political use of their discoveries. This commitment did not involve the transition from one world to another, from the limpid stream of science to the troubled waters of politics, but a mingling of two worlds whose criteria and values, even more than their interests, had hitherto been separate; technical competence, grounded in the methodical examination of facts and the objective pursuit of truth, was associated with the dubious combats and conflicts of political conviction. On the side of power, science was recognised not only as a decisive factor in the national 'potential' but also as an essential tool for the exercise of government itself. The scientist found his way to the capitals—he 'goes to Washington' and elsewhere—as one functionary among others consulted by his government or called upon to represent it at international meetings; the process of political decision could no longer do without his expertise.

The extent of this change can well be measured by reference to Don K. Price's analysis of relations between science and politics in the United States—'the nation that was born of the first effort in history to marry scientific and political ideas'.[43] The Founding Fathers of the American nation saw a correspondence between the constitutional system of checks and balances and Newton's system of mechanics, and the leaders among them, such as Franklin, Washington and Jefferson, were scientists or engineers as much as politicians. But in the first place none of them, however gifted at pursuing several careers at once, would have dreamed of confusing his interest in a scientific career, the time spent on it and the skills it brought out in him, with his functions as a statesman, his commitments as a politician or, finally, the type of knowledge he could bring to bear on public action. Furthermore, if Newtonian mechanics offered the constitutional system the model of equilibrium between opposing forces, it was quite certainly to affirm the separation and not the union between the sphere of public affairs and the sphere of private affairs, between the organs of political power and economic interests—the laws of the market, in effect, acting as the law of gravity and the goodwill of politicians as the prime mover.

42. See, in particular, Alice Kimball Smith, *A Peril and a Hope: The Scientists' Movement in America, 1945–1947*, the University of Chicago Press (1965), 528.
43. Don K. Price, *The Scientific Estate*, The Belknap Press of Harvard University (1965), 5.

Today, in contrast, these frontiers are increasingly blurred, by very reason of the development of scientific research and the importance it assumes for the military, diplomatic, industrial and commercial potential of nations. The American government intervenes directly in the activities of the private sphere, to such an extent that the Founding Fathers, if they came back to earth, would imagine that the constitutional system had been overthrown, or that Newton had been disproved, or—why not?—that the prime mover had ceased to function. In 1969 nearly four-fifths of the 16 billion dollars spent by industry on research and development came from the Federal budget; in less than ten years, since the challenge of the first Sputnik, while research expenditure by the private sector itself has doubled, the total flow of Federal funds to industry for this purpose has gone up more than fourfold.

The development of this partnership between the government and private interests—the universities as well as industry—is not purely the consequence of the expansion of the research system; it is also, and indeed mainly, due to the nature of the results produced by that system. The second world war gave rise to a great deal of research which was all on an unprecedented scale, such as radar, DDT, operational research and the like. But the perfection of the atom bomb led to a mobilisation of researchers, to expenditure and to an association between the state, the universities and industry which was different from other military research in more respects than in its mere scale. With its 15,000 scientists and engineers, its 300,000 technicians and operatives, its cost of more than two billion dollars, the Manhattan Project might well have been regarded in 1945 as the biggest research project ever conducted, but it was the result of the enterprise and not its scale that was decisive.

The perfection of the atom bomb demonstrated that the time-lag between theoretical research and practical applications could be prodigiously shortened, if people were prepared to pay the price in men, money and logistics; but above all, it introduced into the balance of power a completely new type of weapon with completely new implications. There is a very close link, Don K. Price has very shrewdly observed, between the progressive disappearance of the frontiers which divide the different scientific disciplines, especially the frontier which divides pure research from applications, and the disappearance of the frontiers between the government sector and the private sector, between political interests and economic interests, between the field of the politician and the field of the scientist. But if this overlap is henceforth inevitable, it is primarily because another frontier has been swept away by atomic research—the frontier between peace and war.[44]

Until then, military research had been content to adapt civil technologies to the needs of war, thus involving no radical innovation either in science or in politics. During the second world war, scientific research was used for the first time as a source of new technologies whose influence was to be no less decisive on the post-war period than on the termination of hostilities. After that, political power could no longer leave science to itself but, on the contrary, had to force the pace of discovery and innovation. Researchers in the United States were mobilised 'on the spot'; research, in other words, instead of being located purely in public arsenals and laboratories, was centred in the universities and private industry.[45] The demobilisation of researchers at the end of the war did not put an end to the mobilisation of science. For those very scientists who lived in the 'laboratory-barracks' of Los Alamos, the return to their home university might well have meant the return to their peaceful tasks of teaching and research; but those tasks were to take not so much a new form as a new orientation. In fact the alliance between government, industry and the universities was not only kept in being but was made even closer. The balance of power determined by atomic weapons meant that in any new world war the issue would no longer turn on the ability to convert a peace economy and increase mass production but on an immediately 'realisable' scientific and technical capital. The mobilisation of science must therefore be the subject of permanent arrangements, organised and controlled by the government. Science was lodged in the heart of politics and could no longer escape.

44. *Ibid.*, 24, 39 and 59.
45. *National Science Policies: United States*, OECD, Paris (1968), 35.

CHAPTER 3

Knowledge and Power

'You have raised the curtain on vistas of a new world': those were the words with which General Groves, the military commander of the Manhattan Project, of which Oppenheimer was the scientific director, took leave of his former 'troops' of researchers when he handed over his powers to the civilian authorities.[1] Although he was still thinking only of nuclear research, his words were prophetic; the relations between science and power had changed so radically that a new definition of industrial societies was called for. The process of industrialisation could no longer be explained *à la* Marx, by the concentration of industry, or even *à la* Comte, by the scientific rationalisation of labour. The manipulation of intellectual forces was now added to the manipulation of natural forces as the distinguishing characteristic of industrially advanced societies.

The development of science policies in modern societies warrants the description of their common pattern as scientific rather than industrial. The two terms are not equivalent, even if they connote phenomena which are historically and logically complementary. If we can speak today of a 'post-industrial society' it is because the transition from an economy of production to an economy of innovation changes the conditions of the development process rather than merely prolonging them.[2] Now, this change is the direct result of the close relations which have been established between science and the public authorities. The convergence of their interests has set off a massive production of new knowledge and technologies of which the advanced societies make deliberate use: the quantitative change is reflected in a qualitative change.

The success of the Manhattan Project did more than merely make nuclear research a 'vogue' field to which public funds would be assigned with exceptional unstinginess since it promised wonders overnight.

1. R. G. Hewlett and O. E. Anderson, *The New World: A History of the United States Atomic Energy Commission*, vol. 1, 655.
2. See Daniel Bell, 'The Post-Industrial Society', in *Technology and Social Change* (ed. E. Ginsberg), Columbia University Press, New York (1964).

Little by little it transmitted its promise of rapid applications to all
fields of scientific research, inducing the state to take a close interest in
fundamental research, to recognise itself as the principal patron, to
subsidise research for its own sake even where the cost was not offset by
any prospect of an early return. By a sort of collective profession of
faith, scientific activities were suddenly proclaimed a major investment
which no modern society could afford to forgo: good in itself, if not
better than anything else, capable in any event of directly changing the
balance of power between nations and of contributing, more or less
directly, to the attainment of targets of economic growth.

The effort of formulating and carrying out a policy in the sphere of
science marked a new stage in the destiny of modern societies; it was
an affirmation that the pursuit of knowledge had become one of their
essential functions and that they aspired to discharge this function
scientifically. The 'invention of invention', which Whitehead had seen
as the greatest discovery of the nineteenth century, was still only a
starting point.[3] There had arrived the systematic organisation of the
phenomenon, as though scientific discovery and technical innovation for
their own sakes had become the ultimate aim of the process of industrialisation. *In this sense, societies which multiply discoveries and innovations in response to pressures, motives and concerns other than those
of the economy of need are scientific—or post-industrial.*

The novelty is expressed not only by the size of the total flow of resources to research activities (from 1 per cent to more than 3 per cent
of gross national product for most of the industrialised countries), nor
by the rate of growth of those resources, faster and often very much
faster than the growth rate of the gross national product (since 1940
they have practically doubled every five years in the United States, and,
since 1950, the same growth rate has been attained in most of the
industrialised countries). In fact, the novelty lies in the general attitude
of society towards research activities. Science is no longer applied, in
Comte's sense, to the organisation of production, but *society itself is
organised with a view to scientific production.* This reversal is illustrated by the growing importance, both in numbers and in capital, of
the 'science-based industries' and the role they play in the modern
economy by substituting competition through innovation for competition through production. But it is also illustrated, on a different scale, by
the idea we can form of what a new world conflict would be like when

3. A. N. Whitehead, *Science and the Modern World*, reissued by the New
American Library, New York (1953), 98.

the capacity of strategic response would for the first time be defined not by the mobilisation of economic resources, but by the command of knowledge and technologies immediately applicable on the almost instantaneous transition from a state of peace to a state of war.

Whereas for centuries scientists have gone doggedly about their business with little repercussion on world affairs, now not only are world affairs bound up with the scientists' work but collective choices are resolutely made which affect research activities, that is to say, that part of human labour which is not unreasonably deemed one of the least predictable and the least easy to measure in terms of economic return. And yet it is this very argument of economic return which is most frequently invoked to justify society's growing commitments to scientific research—and not solely for the purpose of persuading the legislative or executive authorities that they should not stint their research grants. Discussions on the validity of the research effort as a collective objective thus appear as the extension given by modern societies to Aristotle's definition of man by his 'desire for knowledge'. But in Aristotle's day this formula did not involve the commitment of large resources; the production tasks essential to the society of the day were in no way affected by the development of scientific knowledge or by the cost of that development. In our days even *theoria* demands considerable resources (without particle accelerators there would be no progress in knowledge of the structure of matter) which are in danger of being diverted from other purposes 'useful' to the well-being of society in a different way or more directly.

The priorities of power

We can list the reasons why the state sees fit to encourage research activities in the following scale of priority, in diminishing order of the proportion of resources nowadays allocated to the different fields: military objectives come first, followed by reasons of prestige, economic motives, social objectives and, finally, the advancement of science for its own sake.[4] Science policy is historically the child of war and not of peace. It is not only because the idea of science policy is very recent that it cannot be isolated from the historical context in which it originated and developed; it is above all because this context has not been

4. On this order of priorities, see *Problems of Science Policy*, OECD, Paris (1968): J.-J. Salomon, 10, C. Freeman, 55–6, J. Spaey, 127–8; and *International Statistical Year for Research and Development*, vol. 1, *The Overall Level and Structure of R & D Efforts in OECD Countries*, OECD, Paris (1967), 27–31.

s.p.—3*

changed by the type of peace which was the heritage of the last world war. The impossibility of demobilising scientific and technical resources led to the race, first, in atomic weapons, and then in intercontinental missiles, each new weapons system involving a further escalation in perfecting systems of detection and deterrents.

For the countries already members of the 'atomic club' or candidates for it, the strategy inherited from the last world war means that it is no longer enough to avoid being at the mercy of an adversary; above all, he must be forestalled in the mastery of new weapons systems. In this competition, limited only by the constantly receding prospect of disarmament, science and technology have become the outpost of diplomatic and strategic commerce, each gain recorded in one camp calling for greater effort in the other, a boosting of the means employed and the technological objectives to be attained, unceasingly escalated precisely by the renewal and extension of technologies. Thus, science policy makes its first appearance as an effect and condition of the climate of insecurity to which rivalries between the 'Great Powers' are condemned. But the countries outside the 'atomic club' equally fail to escape the consequence of this competition. In Europe, after being linked with the lag caused by the war, with the decline in scientific and technical potential, with the limitation of resources and the maladjusted educational and research structures, the atmosphere of urgency in which science policies gathered momentum was prolonged by the need to face the competition waged for innovation and favoured precisely by the technological race between the 'Great Powers'.

It is not surprising that the type of authority created to handle these questions in peacetime should recall in the first place the new organisational structure of a war cabinet, as close as possible to the executive summit. The need to coordinate the different sectors affected by scientific research could, of course, explain why the decision-making bodies are located as high as possible in the administrative hierarchy. But more than this is at stake, since any decision about the orientation of research, even non-military research, has strategic implications. Born in a climate of preoccupation with strategy, science policy, even in widening its field of action, has nowhere managed to escape from that climate. The figures alone are enough to demonstrate this; in 1968 more than 50 per cent of research and development expenditure in the United States was for defence purposes, 45 per cent in France and 40 per cent in Britain. If we add atomic and space research the figures go up to more than 80 per cent for the United States and more than 60 per cent for Britain and France.

This high proportion of expenditure on military, nuclear and space research is found only in the most industrialised countries which are already members of the 'atomic club'. But if we allow for the diffusive effects of military research in the civil field, if only by its influence on the progress of certain applications, the mastery of complex systems and the development of new know-how, the indirect consequences of strategic competition affect even the civil options of countries with no atomic weapons. The outstanding example is the design and construction of supersonic aircraft, nuclear reactors or big computers, which are out of the question for medium-sized countries, and still more so for small countries, without sacrifices which would endanger the balance of economic development or would result in irreversible specialisation. Furthermore, government procedures for sharing the costs and risks of military research, especially development contracts, are being increasingly applied to civilian research in countries which have no big military research programmes as much as in those which have. The military mobilisation of science seems the ideal model of the action to be taken to develop both the organisation of research and the organised exploitation of its results.

The industrial society is characteristically a mobilised society concentrating its efforts on a few objectives, among which defence and technical training and development absorb the bulk of the resources. Just as war follows the image of politics, so the economy has adopted the image of war: if '. . . war is . . . a continuation of political intercourse, a carrying out of the same by other means',[5] the means of the economy are a continuation into peacetime of the means of war. No doubt there is an appreciable difference: the adversaries always try to dictate to each other, but it does not follow that '. . . a reciprocal action results which in theory can have no limit'.[6] The clash of major interests is not always settled by bloodshed, or at least recourse to extremes is effected either through the medium of other countries, or on a sort of playing field which excludes violence as such. Technological competition between the most industrialised countries very closely fits Huizinga's definition of war as having a play-quality, a ludic and agonistic function, a sort of game in which each side tries to beat the other for the glory of victory rather than for the advantage of conquest.[7] So long as there is no recourse to the *ultima ratio*, it shifts on to the playing

5. K. von Clausewitz, *On War*, Random House, New York (1943), 16.
6. *Ibid.*, 5.
7. J. Huizinga, *Homo Ludens: A Study of the Play-Element in Culture* (trans. R. F. C. Hull), Routledge and Kegan Paul, London (1949), 89–90.

field the element of brute force which constitutes the reality of war.

Like children, in effect, the 'big boys' play at frightening each other, but this game does not rule out the threat in earnest illustrated by the investments made to wipe out the adversary's real or imaginary lead or the reality of the economic advantages flowing from these investments. It takes the challenge of a gap—atom gap, missile gap, technological gap—to perpetuate everywhere both the process of state commitment to research activities and the process of speeding up the transfer of the applications of science to the economic circuit. Modern war, treated as a game, finds its sublimated form in scientific research, itself full of hazards.[8] If we suddenly reverted to peace of the traditional type—to a 'conventional' peace, in the sense in which we speak of 'conventional' weapons, that is to say, the lessening of tensions by the return to an economy which no longer tended to simulate war by competition —the sanction given to scientific and technical investments would lose a great deal of its justification. As Galbraith very rightly says, 'Obsolescence in a technological competition is a nearly perfect substitute for battlefield attrition'.[9]

Nearly perfect; where wartime attrition might generate the opposition or even the insurrection of the mobilised populations exposed to the perils of the battlefield, the risk of obsolescence, in contrast, leads to the constantly closer association of mobilised science with the aims of the industrial system. And the revolts which may result are not of a size to compel a change of orientation: first, because the protesting scientists do not constitute a group which is sufficiently homogeneous, numerous and well organised to carry the same weight with the political authorities as a rebellious party or a mutinous army; second, and above all, because radical rejection of the priorities set by the state would mean surrendering the grants whose growth is conditioned by the continued observance of those priorities even when the grants are not earmarked for priority fields. Just as research as a whole would not have developed so far or so fast without the impetus given by the last world war, so all scientific disciplines from the sciences of nature to the sciences of man end by benefiting from the priorities rightly or wrongly assigned to some of them. The undeniable gap between the financing of the natural sciences and the human sciences, or even within the natural sciences between those which show an immediate return and those whose appli-

8. For Clausewitz, war is '. . . of all branches of human activity, the most like a game of cards'. (*On War, op. cit.,* 15).

9. J. K. Galbraith, *The New Industrial State,* Houghton Mifflin, Boston (1967), 330.

cations are indeterminate, should not be allowed to mask the equally undeniable fact that all research sectors have found their resources increasing over the last decade. State patronage, only yesterday limited to certain fields, has become an industrial undertaking, spreading step by step to all sectors of science.

The second category of objectives cannot really be distinguished from the first, for reasons of prestige extend into technological competition the military reasons which affirm the relative strength of nations. As Raymond Aron has recalled, the series of 'eternal objectives' pursued in international relations is complete only if the third term of glory is added to security and strength. In one sense, 'glory is merely another name or another aspect of power; it is, as it were, power recognised by others, whose renown spreads throughout the world'.[10] While, in the abstract, this division draws a distinction between three specific concepts, in practice they are inseparable. Just as a firm which brings an innovation into the market can dictate to its competitors, so the acquisition of priorities in discoveries, no less than technological performance, is a sign of the power of nations and, by implication, of the validity of the political systems they uphold.

Battles for priority have always been part of the history of science, but in the old days they were fought out between individuals, whereas nowadays they have been taken over by the state. Nobel prizes are inscribed among the national honours just like the achievements of Greek athletes at the Olympic Games; the only difference is that the laurels of science and technology no longer help towards a truce to rivalries, but rather prolong them. No example is more revealing than that of space competition, the very type of 'competitive struggle', in the words of François Perroux, in which technological achievements are chalked up as proof of the simultaneous realisation of the three 'eternal objectives'. In the first place, space competition is never free from military undertones, since the orbiting of spacecraft or moon landings may always guarantee a further strategic advance.[11] Secondly, the prestige accruing to the victors in the different stages in the race—the first sputnik, the first man in space, the first woman in space, the first man around the moon, the first man on the moon and so forth—underlines

10. R. Aron, *Peace and War—A Theory of International Relations*, New York (1966).
11. The inextricable mixture of military motives and prestige is explicit in the documents which led up to the creation of NASA, whose function is nevertheless essentially civilian. See Vernon Van Dyke, *Pride and Power: The Rationale of the Space Research*, University of Illinois Press (1964).

the strength of technical means which are themselves part of the military potential.[12] Finally, the shift of violence between the great powers on to the ludic and agonistic field brings all the points scored by the adversaries on the political, military and technological planes into the same complex of challenges.

Economic motives, whether they concern competitiveness or growth, are theoretically independent of the other two. They are so in practice when they are expressed by industrial research conducted by private firms without government support; but as soon as government support comes in, the line of demarcation between economic concerns and military concerns ceases to be sharply drawn. On the one hand, international comparison of research efforts tends to indicate that the financial support of research for 'economic objectives' is inversely proportional to the financial support of research for military objectives.[13] On the other hand, it is clear that the military research effort itself acts as an economic stimulus, directly through the new products or processes which it introduces to the market, and indirectly by the dissemination of knowledge and the rise in the general level of professional qualifications. The very conditions of technological competition themselves are largely defined by the progress made by private firms in the field of military research. Thus it was military objectives that created from scratch the field of the great electronic computers which spread to the private sector by the general introduction not only of the use of these new machines but also of the application of entirely new methods of design and management and production techniques.

The problems of the economic effects of expenditure on military and space research are the subject of debate which is still inconclusive. Some writers stress its advantages for the civil economy, others the drain it involves, to the detriment of the equilibrium of firms and

12. The first sputnik was such a shock to American opinion that it induced Congress and the President to create most of the institutions (up to 1973) responsible for science policy. The literature on the political and ideological use of space exploits falls into the realm of pure fantasy by the number and nature of comments made, ranging from the press conference at which Gagarin declared he had not met God in space to the statements of some American politicians such as the following: 'If the USSR can use women as well as men [in space exploration] and we use only men, they of course would have a tremendous man/woman power advantage over us.' (Senator Stuart Symington, cited by H. L. Nieburg, *In the Name of Science*, Quadrangle Books, Chicago (1966), 10).

13. According to the OECD definitions, 'economically motivated R & D' is the sum of commercial and industrial R & D, R & D for the economic infrastructure, and agricultural R & D. (OECD, *op. cit.*, note 4, vol. 1, 28.)

universities. The debate is inconclusive because there is no scale by which to measure the relative weight of these different effects on the economy. It may well be, moreover, that the spin-off of military and space research as it affects the civil economy will be less important in future than it has been in the past twenty-five years, since the types of technological processes and innovations generated by this research are becoming so complex and so refined that they can no longer be used in industrial production; the excess of science is, in a sense, restoring the situation before the second world war, when the relative absence of science from military programmes led to the transfer on to the battle-field of instruments already known and used by the civilian sector rather than the introduction of new technologies from the battlefield into the civil sector.

One thing is nevertheless manifest: however immeasurable the in-fluence of military research on the civil economy, that economy would not be so marked by technological innovation as it is today without the spur of military programmes. If war, 'the midwife of History', is one of the forms of commerce between nations, the industrial system has become such that it can give this commerce all the forms of war, from open conflict to battles for the conquest of markets, from armed rivalry to technological competition. It is in any event doubtful whether government support for research would have assumed the magnitude it did under the pressure of military motives if it had been directed towards purely economic objectives. The countries which spend more on research for economic purposes than on military research are, with the exception of Japan, also those whose total research expenditure is the smallest percentage of gross national product.[14]

These reservations naturally do not mean that a reversal is neither de-sirable nor possible, or, in other words, that more military research and less research for economic purposes is needed. In the small countries, and in all countries among small firms, the state already supports research for economic purposes to make it possible to face competition from lar-ger firms or from countries which have the advantage of economies of scale. Even so, it must be assumed that the modern industrial system, without military motives, still desires the acceleration of technical change and is prepared to accept the same measure of the risks inherent in re-search. The convergence of military and economic objectives justifies the state in taking over the financing of programmes which are too risky to be acceptable to private industry, but the funds which might be released

14. *Ibid.*, 29.

by cutting down military programmes will not necessarily be appropriated to research for economic purposes. As Galbraith very rightly says, 'while all expenditures, whether for arms or old age pensions, add to demand, not all play the same role in underwriting technology.[15]

This limit is even more apparent for the social objectives which are the poor relation in most national science policies. Social objectives are interpreted as meaning health, hygiene and the abatement of nuisances (air and water pollution, problems of urban development, etc.). The financial and human resources allocated to research in these fields mobilise only a small proportion, both in absolute and in relative terms, of the total research effort.[16] These resources could—and should—be augmented; greater investment in medical research would result beyond all doubt in immeasurable progress in the prevention and cure of disease.

At the same time, there is a sort of naïvety or gullibility—let us call it 'scientistic illusion'—in the belief that problems of health and, more generally, of the environment could be solved in proportion to the expenditure on scientific research towards their solution. The very successes gained by technology in the last quarter of a century—or uneasy conscience about those successes won through military research programmes—induced people to think (or to profess) that the same effort devoted to social objectives could be expected to yield a similar return to that of nuclear or space technology. Hence all the talk about science applied to the problems of underdevelopment or to curing the traumas produced precisely by the ever-growing pace of urban civilisation and technical change. Thus it was with the 'Great Society' programme in the United States, a large part of which was designed to orient or re-orient research and development towards improving the framework and conditions of social life; the list of problems which science was to help in solving ranged from oceanography to crime and poverty in the cities, by way of weather and climate control, the abatement of air and water pollution, and the organisation of transport systems.[17]

15. Galbraith, *op. cit.*, note 9, 339.
16. The OECD definitions add to 'socially-oriented R & D' (health, hygiene, research into arid and underdeveloped zones) a group of 'miscellaneous R & D' which includes 'all intramural government expenditures which could not be allocated elsewhere', that is to say, a significant proportion of expenditure on non-directed basic research. The data thus compiled do not allow any reliable comparison, but they do give a relative idea of the insignificance of the resources devoted to this sector in comparison with others. (OECD, *op cit.*, note 4, vol. 1, 30.)
17. *Reviews of National Science Policy: United States*, OECD, Paris (1968), chapter 14, 'Science and the Great Society', 289–303.

Among these problems, the ones which most obviously depend on scientific and technical research and to which, moreover, the greatest resources have been allocated are also the ones which have military aspects, namely oceanography, meteorology and the mastery of climate, and transport systems. Apart from strategic considerations, these fields are at the same time the source of new markets and therefore constitute a new stake in economic competition:

> [The food-from-the-sea programme] offers a new opportunity for the United States to provide world leadership through a long-range programme to exploit the oceans as a relatively untapped source of protein for the undernourished.[18]

Vice-President Humphrey, then responsible for coordinating programmes in this sphere, was not slow to point out that oceanography offered more chances of profit than any other scientific enterprise.[19] Whether these words are justified or not, the same economic and strategic considerations that induced the United States to create the National Council on Marine Resources and Engineering Development constrained other countries to set up similar institutions;[20] one of the characteristics of technological competition is this phenomenon of imitation or propagation in the smaller countries of the priorities set by the bigger countries and the measures they have taken to put them into effect.

As for the other problems—poverty, crime, racial conflicts—it is easy to see what part the social sciences might perhaps play, at any rate to some extent, but what is the relevance of the natural sciences? Just as the prowess of technology generates the belief that science is all-powerful, so the success of the major government-sponsored priorities makes people think that science policy is all-embracing. Even if social objectives were to become the top priorities of power, it would be surprising if science and technology could solve this kind of problem. The great utopias of the past were social or political: science entered into them as an instrument of progress in so far as its own ends coincided with or expressed those of the dream society. Nowadays novels of the future merely envisage technological progress as a means whereby political and social organisation will attain its own ends. In taking the place of utopias, science fiction does not aim at liberating man from the tyrannies of superstition or ignorance, or organising a more human world for

18. *Marine Science Affairs, A Year of Transition, Ibid.,* 309.
19. *Ibid.,* 310.
20. In France, the Centre national d'Exploitation des Océans (CNEXO) was created in 1967.

mankind; it merely subordinates the fate of man and the world to the progress of technical conquest. The myths of human 'progress' have been killed by the progress of science itself; the 'scientistic illusion' regards technology as a substitute for social and political choices.

Knowledge as the objective of power

If support for research becomes a state affair, it is because political power identifies its own cause with the cause of science; the advancement of science for its own sake has become one of its objectives, even though it may have lowest priority. And this objective, once again, stems from an instrumental conception of science. As Christopher Freeman says:

> ... the fact that fundamental research leads to practical economic results by rather a roundabout route does not mean that the motive for supporting it is not utilitarian. This is quite clearly shown by the example of the Soviet Union. The Soviet Union, from 1917 onwards, categorically rejected the whole idea of science for its own sake, or pure science, just as they rejected the theory of art for art's sake, but nevertheless Soviet science policy has always found it very well worth while to allocate substantial funds to fundamental research for purely utilitarian reasons.[21]

But the United States provides an equally striking example in so far as it has always been the policy even of the most specialised agencies, such as the Department of Defense, NASA and the Atomic Energy Commission, to spread their appropriations over the whole range of research.

The studies made at the RAND Corporation by the team of economists centred on Burton Klein provided a veritable doctrinal backing for this policy. If the government has to allocate a substantial proportion of military research expenditure to activities not directly related to the perfection of weapons systems, it is because this is the cheapest way of ensuring its adaptability to new strategic conditions in the shortest possible time; the more varied the bill of fare in research, the more it suits all kinds of diet.[22] The fact that the Russians do it without a word

21. 'Science and Economy at the National Level,' in *Problems of Science Policy*, OECD, Paris (1968), 56. Freeman refers in particular to the paper by Jacob Schmookler, *Catastrophe and Utilitarianism in the Development of Basic Science*, taking the view that, ever since the beginnings of modern science, there has never been any genuinely disinterested state support of research activities (in *Economics of Research and Development* (ed. R. A. Tybout), Ohio State University Press (1965), 19–33).

22. B. Klein, 'The Decision-Making Problem in Research and Development', in *The Rate and Direction of Inventive Activity*, Princeton University Press (1962), and 'Policy Issues in Military Development Programs', in *Economics of Research and Development, op. cit.*

and the Americans do it while proclaiming their acts (and the other industrialised countries often hint at it) does not change the obvious convergence of science policies, all of which tend to adapt the research
system to the requirements of the state in the same way.

But nothing has yet been said about state objectives determined by
utilitarian motives. Investment in pure research is a guarantee for the
future on the national scale just as much as on the industrial scale. It is
immaterial that the purpose of the state is wider than that of companies
such as Bell or Du Pont de Nemours, for it is not on the face of it very
different; pure research as a source of innovation is a guarantee against
rival firms—a guarantee of profit and of survival. The fact remains that
an industrial firm, however remote from its short-term objectives may
be the research which it subsidises, never ventures completely into the
unknown. When Du Pont called in Carothers and gave him the means of
conducting his work on polymers freely and without any time limit,
there was nevertheless a presumption that the result of this pure research would one day lead to an application relevant to the company's
traditional products; similarly, the choice made by Bell Laboratories
in endowing with immense resources the research of Shockley and his
colleagues on solid state physics was far from being irrelevant to their
field; just as nylon originated from a chemical industry which already
had experience of man-made textiles, the transistor was perfected by an
industry specialising in telecommunications.[23] The state, for its part, has
no specialised activity which justifies it in producing pure knowledge 'in
general'. However questionable it may be, the order of priorities it sets
for its science policy follows a certain logic; but what is the logic of the
support it gives to activities which appear to be an end in themselves
and nothing more? For example, the study of high-energy particles
or extragalactic systems can have no utilitarian motive unless possible
applications can be envisaged in some indeterminate future, and there
are very few people today, even among the specialists, who would venture on such a speculation.

It is at this level that we must ask to what extent the objectives of
science coincide with those of the state. *The field of pure research constitutes the extreme case of coincidence between the interests of power
and the interests of knowledge; it provides the best illustration both
of the nature of the collective choice of which science is now the sub-*

23. See Richard A. Nelson, 'The Link between Science and Invention: The
Case of the Transistor', in *The Rate and Direction of Inventive Activity, op.
cit.*, 549–83.

*ject and of the impossibility of finding a fully rational justification for
it.* If the reasons which science attributes to the state and which it itself
invokes in soliciting state support were the state's real reasons for
granting it, there would be little ambiguity in the relations between
knowledge and power. But pure science does not express its aims in
terms of utility, and yet the state assumes responsibility for them as
one function among others in the light of its own specific aims.

The question is so far from answering itself that a Committee of
Congress put it to scientists in the country where public financing of
pure research is larger, both in absolute and relative terms, and has
grown faster, than in any other sector of the national research effort.[24]
If we add that, in theory, the United States Federal Government does
not intervene in the university sector, education and research being a
matter for the individual states or the private sector, the point becomes
all the more significant. The approximation between the sphere of
private interests and the sphere of public interests, brought about by the
very development of the research system and the attention the state
pays to it, raises not only the legal question of 'conflicts of interest' but
also the more general question of the subordination of the ends of one
sphere to the ends of another.

None of the scientists consulted, says the report, 'challenged the proposition that the purposes of government, as opposed to the techniques
of government, are non-scientific': the questions put to them 'must be
answered in terms that generally lie outside science'.[25] The reason is
that if the pursuit of science appears as an essential function of society,
there is no substitute for the alliance between the state and pure re-

24. *Basic Research and National Goals: A Report to the Committee on
Science and Astronautics, U.S. House of Representatives by the National
Academy of Sciences, Washington* (March 1965). In fact, two questions were
asked: 'I. What level of Federal support is needed to maintain for the United
States a position of leadership through basic research in the advancement of
science and technology and their economic, cultural and military applications?
II. What judgment can be reached on the balance of support now being given
by the Federal Government to various fields of scientific endeavour, and on
adjustments that should be considered, either within existing levels of overall
support or under conditions of increased or decreased overall support?' (page
1.) The two questions were dealt with in a series of fifteen individual contributions by the members of an *ad hoc* panel of the Committee on Science and
Policy of the National Academy of Sciences. Most of the contributions interpreted the two questions as follows: (a) how should resources be distributed
between science and the other activities of our society? (b) how should the resources allocated to science be distributed within science?
25. *Ibid.*, 5.

search; modern science is so fashioned that the cost of each new item of knowledge is far higher than the cumulative cost of most of the knowledge acquired a century ago, and modern societies are so fashioned that they are nourished by this knowledge even in their everyday functioning. In a market economy it is the responsibility of the state to support an activity whose return is too indeterminate to attract the support of private enterprise; investment in pure science corresponds in a way to the overheads of society, and its contribution to social ends—knowledge and understanding—is so diffuse and takes such devious routes that only the government can bear the cost:

> It is in the very nature of an overhead that a nice calculation of the 'right' amount to spend on it is difficult.[26]

The theoretical problem of determining this amount turns on an impossible calculation since it would have to be shown

> that basic research yields a social return over its cost that exceeds the return on alternative types of investment of resources.[27]

It is quite clear that, whatever the amount, it would be insufficient in the absence of government support.

The convergence of the interests of knowledge and power is inherent both in the nature of modern science—knowledge is power—and in the objectives of the state—power is knowledge—from the moment that the debate is couched in the terms of the scientific society. Thus, all the contributions to this study— except one, by an economist, Harry G. Johnson—agree in identifying, as is done in particular by Harvey Brooks, four goals of society to which basic research contributes and which justify its support by the state:

> Basic science, *per se*, contributes to culture; it contributes to our social well-being, including national defence and public health; to our economic well-being; and it is an essential element of the education not only of scientists but also of the population as a whole. In deciding how much science the society needs, one must decide how the support of science bears on these other, politically defined, goals of the society.[28]

However, as soon as we try to delve deep into the nature of these goals, it becomes apparent that none of them lends itself to rigorous demonstration; value judgments inevitably outnumber findings of fact and even further outnumber unchallengeable conclusions. This is not only

26. Carl Kaysen, 'Federal Support of Basic Research,' *ibid.*, 153.
27. Harry G. Johnson, 'Federal Support of Basic Research: Some Economic Issues', *ibid.*, 135.
28. *Ibid.*, Summary, 5.

because any discussion of the goals of society sparks off an ideological debate, it is also because in this debate the postulates 'it is just as if . . .' or 'all other things being equal . . .' refer both to such a multitude of heterogeneous facts that it is idle to look for any rational link between them, and to an irreversible evolution of the research system which has its repercussion on the whole definition and orientations of modern society.

It is significant that in this anthology, the only contribution to cast doubt on the inevitable link between pure research and society is precisely the one that questions the whole goals of society:

> Clearly [says Harry Johnson] if the public is convinced that a scientific culture is desirable, it is perfectly appropriate for the taxpayers' money to be used to support scientists and scientific research. But to the extent that scientific activity is of the character of a consumption good . . . its claims for public support need to be weighed against other pressing claims on the social surplus, such as the relief of poverty, the mitigation of social problems, the needs of the less-developed countries. . . .[29]

Not that there is even any reason to believe that effective action to meet these last-named needs depends on greater efforts in the matter of pure research:

> . . . money applied with existing knowledge would suffice, because it is the nature of our political and social attitudes and institutions, not the backwardness of our social scientific knowledge, that is primarily responsible for the problems. With respect to poverty, for example, a major obstacle to more effective policies is not lack of knowledge of what causes poverty, but the belief that poverty is the poor person's own fault and that giving him money will sap his initiative.[30]

The scientific society is not only the society which wills science as one of its ends, but also the society in which scientists desire their own ends to coincide with those of society. Of the four goals accepted in the discussion—culture, social well-being, economic well-being (including national defence) and education—there is none, not even the cultural goal, which does not serve to demonstrate this will to identification.

> Basic scientific research [says Harvey Brooks] is recognised as one of the characteristic expressions of the highest aspirations of modern man. It bears much the same relation to contemporary civilisation that the great artistic and philosophical creations of the Greeks did to theirs, or the great cathedrals did to mediaeval Europe.[31]

But cannot art, music, literature and the dance also be said to be equally characteristic expressions of modern man? And why research in the

29. H. G. Johnson, *ibid.*, 132.   30. *Ibid.*, 133.
31. H. Brooks, 'Future Needs for the Support of Basic Research', *ibid.*, 84–5.

natural sciences rather than in the humanities? If science is recognised as eligible for government support where other forms of cultural activities are not, responds the same author, it is because it is a cultural activity which is generally believed to transcend private value-systems. The reasoning is such that it must be cited in full:

> The only definite answer that can be given to these questions lies in the nature of science as a system of acquiring and validating knowledge. Science—especially natural science—has a public character that is still lacking in other forms of knowledge. The results of scientific research have to stand the scrutiny of a large and critical scientific community, and after a time those that stand the test tend to be accepted by all literate mankind. Outside the scientific community itself this acceptance tends to be validated by the practical results of science. If it works it must be true. There is no question that the successful achievement of an atom bomb provided a certain intellectual validation for nuclear physics, quite apart from its practical value. Part of the public character of science results from the fact that it is always in principle subject to independent validation or verification. It is like paper money that can always be exchanged for gold or silver on demand. Just because everybody believes that he can get gold for paper, nobody tries; so the public seldom questions the finding of science, just because it believes that they can always be questioned and validated on demand. This is much less true of other forms of knowledge and culture, which may be of equal social importance but are more subjective and more dependent on the vagaries of private tastes and value systems.[32]

Nothing demonstrates more clearly than this passage how the link between knowledge and power tends to identify the ends of each of them by the social recognition of scientific truth: *the operational character of modern science turns its theoretical procedure into a political procedure.* Even if it is recognised that the results of scientific research, unlike other forms of knowledge or culture, have this public aspect which 'transcends private value systems', it is difficult to assert—and Harvey Brooks admits it—that these other forms of knowledge have less social importance and are therefore less worthy of public support (even in the United States, where cultural activities are largely supported by private patrons or foundations). Furthermore, if we speak of pure research in the strict sense—research which does not lend itself to any foreseeable application—we may question the extent of the 'public character' of science; it is, after all, an esoteric enterprise, whose content, interest and values can be appreciated only by its devotees. However big the army of researchers, it is still smaller than that of music-lovers or art-lovers.

At a deeper level, if one regards the works of science as the symbol

32. *Ibid.*, 85–6.

of the twentieth century, in the same way that the Pyramids are the symbol of Pharaonic Egypt, the cathedrals the symbols of the Middle Ages or the Palace of Versailles the symbol of the age of Louis XIV, and so forth, one may also look upon them as monumental undertakings whose price was not necessarily indispensable to the survival of the civilisations they express. According to Herodotus, the Pharaoh Cheops employed an army of a hundred thousand men for twenty years in constructing his pyramid and sold his daughter to buy stones, and the enterprise left Egypt devastated as though by war.[33] The real difference, which should make us think, between these gigantic enterprises of the past and the gigantism of modern scientific research, is that the former came to an end, while the latter, even if it grows—and its cost with it— exponentially, has no conceivable term except in the event of a nuclear apocalypse. But, precisely, paper money has a market value only so long as there is no insolvency.

Seen from its cultural aspect as an end in itself, pure science has no greater claim to state support than any other cultural activity; it must have the backing of national goals before it becomes a collective adventure. In other words, science is an end of society only when it provides a means, If the aims of the state were different, science might well have a different status:

> Insistence on the obligation of society to support the pursuit of scientific knowledge for its own sake differs little from the historically earlier insistence on the obligation of society to support the pursuit of religious truth, an obligation recompensed by a similarly unspecified and problematical pay-off in the distant future.[34]

From this point of view, neither the cultural argument nor the educational argument can be dissociated from the context of technological competition in which the state adapts the research system to its ends. If more science is needed, more researchers are needed, and since the process which leads from basic research to development and production is continuous, the training given by pure research is an essential element in the general system of education required by the scientific society.

33. This argument was put forward in an address by Lord Bowden, formerly British Minister of Science, published in *New Scientist*, London (30 September 1965), and in his opening address, *Problems of Science Policy*, OECD, Paris (1968), 21–41. Variants are to be found in the literature on the allocation of resources to science, especially in the studies from the review *Minerva* edited by Edward Shils in *Criteria for Scientific Development: Public Policy and National Goals*, MIT Press, Cambridge, Mass. (1968).

34. H. G. Johnson, in *Basic Research and National Goals*, op. cit., 132.

Whatever the terms in which pure science defines its ends in relation to power, they are instrumental: culture is not consumed, but science is consumed as a technique among others, subordinate to the ends of society. Science is one of the instruments of power, but power is also the means whereby science prospers; the argument of economic return, however evasive it may be, is the only persuasive one. If science is to have state support, its ends must be identical with the ends of society. The growth and influence of the research system meet the needs of the industrial system, just as state priorities meet the needs of science. It took a miracle to enable the builder of the Basilica of Saint-Denis to erect a sumptuous altar:

> The martyred Saints themselves [writes Suger] furnished us unexpectedly with much gold and many very precious stones as though they wanted to tell us with their own voices, Whether you will it or not, we desire to have the best there is.[35]

Since the consummation of the alliance between knowledge and power, the citizens of modern societies have heard nothing but the same language.

The pursuit of science fits in with the ends of power itself, however remote it may be from any purpose outside itself and however remote its own purposes may be from those of the state. 'In a certain sense', says Harvey Brooks, 'it [basic research] not only serves the purposes of our society but *is* one of the purposes of our society.'[36] The destiny of modern societies appears so inextricably interwoven with scientific activities that the mobilisation of researchers must be perpetuated on all the fronts of knowledge; in seeking to organise and develop all their enterprises scientifically, including those such as pure research which by tradition and nature least lend themselves to it, they must rely on a steadily increasing number of researchers in a steadily increasing number of sectors of activity. It is food for thought that in the most highly industrialised societies today researchers outnumber priests and army officers combined—remembering that some army officers, moreover, should also be classified as research workers.[37]

35. Cited by Jean Gimpel, *Les Bâtisseurs de cathédrales*, Editions du Seuil, Paris (1961), 21.

36. 'Future Needs for the Support of Basic Research', in *Basic Research and National Goals, op. cit.*, 85.

37. It is estimated that the total numbers employed in research activities have gone up tenfold in twenty years; in 1965 more than a million people were engaged in research and development in the United States, the same number in the USSR and more than half a million in Western Europe (C. Freeman and A. Young, *The Research and Development Effort in Western Europe, North America and the Soviet Union*, OECD, Paris, 15.)

For the theorist of the nineteenth century, the age of the industrial society was to replace priests and soldiers by industrialists and scientists; but the new world which is the aftermath of the world war has turned the scientists into a new priesthood allied with military power:

> . . . it had begun to seem evident to a great many administrators and politicians that science had become something very close to an *establishment*, in the old and proper sense of that word: a set of institutions supported by tax funds but largely on faith and without direct responsibility to political control.[38]

It is no accident that the theme of scientists defined as the priests of a religion whose cult is bound up with power is constantly recurring in the literature on science policy; the movement which, from the beginnings of modern science, has made scientific research a secularised profession culminates, with science policy, in metamorphosis into a collective faith, into a state religion:

> To an important extent, indeed, scientific research has become the secular religion of materialistic society; and it is somewhat paradoxical that a country whose constitution enforces the strict separation of church and state should have contributed so much public money to the establishment and propagation of scientific messianism.[39]

It is moreover striking that it was a general and a President of the United States who uttered the gravest warning against this conjunction of the military and industrial system with scientific research. In his farewell speech as President, General Eisenhower referred to the risks of a public policy becoming the captive of a scientific and technological élite and of the military–industrial complex to which that élite owes its existence.[40] Whether the priests of the new religion have taken over the power, or whether power has been converted to the desires of the new priesthood, science seems to have become all the more a matter of faith in becoming more closely an affair of state.

38. Don K. Price, *The Scientific Estate*, 12.
39. H. G. Johnson, in *Basic Research and National Goals*, 141, note 4. This theme appears repeatedly in many books, even in their titles, such as Ralph Lapp, *The New Priesthood: The Scientific Elite and the Uses of Power*, Harper and Row, New York (1965), and Spencer Klaw, *The New Brahmins: Scientific Life in America*, Morrow and Company, New York (1968).
40. *New York Times* (22 January 1961). The history of this statement is significant; in the face of the immediate reactions it aroused among scientists, a White House spokesman made it clear that President Eisenhower was not speaking of scientists in general but only of 'scientists allied to military and industrial power'. Dr George B. Kistiakowsky, at that time Special Assistant to the President for Science and Technology, had to publish an 'authorised' interpretation for the benefit of the scientific community (*Science*, no. 3450, 10 February 1961).

PART II

Politics in Science

King Charles X was one day visiting the Polytechnique when his curiosity was aroused by a model of the hyperboloid of one sheet. The Professor tried to explain to the King that this surface of revolution was generated by a straight line. At his wits' end, the Professor (he was called Leroy) finally exclaimed, 'Then Your Majesty will just have to take my word of honour for it!'

BACHELARD

CHAPTER 4

The Scientific Research System

Scientific research pays a price no smaller than other activities pay for its links with the industrial system. Affluence has its counterpart in organisation, programming and planning. Favourite son that it is, scientific research still has to obey the rules of the game laid down by modernity. The quantitative change in research activities since the second world war is reflected in a qualitative change: from being an ideal, a vocation, a culture limited to a few, research has become a profession open to the masses; as the source of innovations which can be rapidly exploited, it forms an integral part of the industrial system. Some researchers regard this change as a betrayal of the ends (and of the interests) of science, in so far as it ought to be concerned solely with the pursuit of truth; the climate of utility in which it has blossomed compromises it, alienates it and, in a word, prostitutes it.

Science is not philosophy, but in this idea of science there is something of philosophy in the sense in which, for Socrates, it was contrasted with the activities of the sophists:

> I consider you indeed to be a just man, Socrates, but by no means a wise one; and you appear to me yourself to be conscious of this; for you ask money from no one for the privilege of associating with you. . . . It is evident, therefore, that if you thought your conversation to be worth anything, you would demand for it no less remuneration than it is worth. You may, accordingly, be a just man, because you deceive nobody from covetousness, but wise you cannot be, as you have no knowledge that is of any value.[1]

The argument of Antiphon, as reported by Xenophon in the *Memorabilia*, amounts to saying that services rendered have no value except what people are prepared to pay for them. Since Socrates does not sell lessons in wisdom, what he teaches has no value—it is beneath calculation or derisory. Regarded as a research activity with no object except the extension of knowledge—or the researcher's own satisfaction in con-

---

1. Xenophon, *Memorabilia*, chapter 6, 11–12; trans. J. S. Watson in *Socratic Discourses*, Everyman's Library, Dent, London (1954), 32.

tributing to it, the joy of discovery—science may pass for an enterprise whose services cannot be evaluated in the absence of a 'fair price'; knowledge, understanding, the extension of our knowledge and understanding of nature, are all 'goods' upon which no value can be placed since they are apparently of interest only to their devotees, whose sole aim is their attainment. Like the teaching of Socrates, science obeys no rule except that of the spirit.

Truth and profit

This parallel must not be carried too far: Socrates did not ply the trade of philosopher. In our days, the 'savants' are paid researchers; in some countries—the United States, Russia—their privileges lose nothing in comparison with those of industrial managers, and their work calls for resources which rival those of other public investments. And yet, even if science has become a profession and now counts as a productive activity, the idea of an 'intellectual good' to be achieved, whose value has no 'fair price', remains—and will always remain—inherent in the scientific approach. The fact that the pursuit of this good demands the interested support of society does not prevent it from seeming a purely intellectual goal; the value of a discovery or a theory cannot be measured by the investments it needs or by the practical results to which it lends itself. A society is therefore conceivable which would be satisfied with its accumulation of material goods and which would devote a growing part of its wealth to the enlargement of knowledge solely because it is a good in itself, or which would even, in so far as our society itself is scientific, affirm that the pursuit of knowledge for its own sake is one of its essential ends. If the state ought not to stint its support for science, it is because of the intellectual goal rather than the practical applications or the services which the state expects from science in the pursuit of its own aims. At best, the aims of the state should not differ from those of science, since its goals should be the well-being or betterment of the body social as a whole and, by definition, the increase of knowledge is one of the most decisive factors in well-being or betterment.

The premises of such reasoning, which was that of Renan in *L'Avenir de la Science*, are repeated in most of the pleading in favour of pure research, and for many scientists they are bound to lead to the same conclusion which Renan intended to draw. If the state owes its support, it is because the state's own interest lies in contributing to the progress of knowledge for its own sake; support should therefore be unconditional. By definition, the guarantee of independence lies in blind support, generously leaving the beneficiaries their complete freedom. Just as the economic theory of 'laissez-faire' relies on freedom of enterprise as the means of attaining the greatest economic efficiency, so the affairs of science should be left to themselves, that is to say, in the hands of scientists, so that knowledge can follow the surest line of progress. Under this conception, the ideal source of finance is the private patron or foundation—the model which, in fact, prevailed up to the nineteenth century.

This debate, on science regarded as an autonomous system faced with political power whose support nevertheless conditions its development, was opened at a time when science alone was really in the position of asking for something, and the political power was, for the most part, indifferent to its requests. In the nineteenth century the fear of undue intervention was all the less securely founded, since the state, far from being a 'partner' in research results and envisaging a strategy to make the best of them, disregarded science as a source of applications. Today the terms have changed and the relation has become reciprocal: the state defines itself, by its own initiative, as a consumer and producer of 'pure' knowledge; it is an interested party in the results of scientific research and, what is more, its interest is paramount—in commercial law it would be called a preferential creditor. The relation of dependence implies the threat of subordination: political power is, by definition, authority and nothing seems more hostile to science than to have to render account to any authority except its own internal discussions. Political power attributes authoritative values:[2] how can science bow to the values of power without denying itself?

The history of science is the history of its struggle against the spirit of authority, whether it be the dogma of the church, the official doctrine of the state or the orthodoxy of a party. From the condemnation of Galileo to that of the Russian geneticists, the secularisation of power has wrought no change in the issue at stake; the scientific enterprise cannot submit to dictation, in either its methods, its content or its conclusions. The limits which Galileo set to the spheres of competence of science and power have not been abolished by the harmonisation of their interests: this intervention with scientists 'would be to order them to see what they do not see, not to understand what they understand and when they seek, to find the opposite of what they find'. The truth

2. See David Easton, *The Political System: An Inquiry into the State of Political Science*, Knopf, New York (1960), 129–46.

which science invokes can be refuted by no other authority save science itself:

> With regard to these propositions and others like them, which are not directly *de fide*, no man doubts that the Sovereign Pontiff always has the absolute power to admit them or condemn them; but it is not within the power of any created being to make them true or false except as they may be so by their nature and *de facto*.[3]

The theme of the autonomy of science is not only a reminder that Socrates' teaching was free; it is also an affirmation of the rights of truth against authority which claims to diminish it, falsify it or subject it to the decrees of orthodoxy.

From this point of view, the intervention of political power in the content and conduct of scientific affairs is not and never will be anything but totalitarianism—the enslavement of the spirit to the convictions of a community. Whether it narrows the field of competence of science, subordinates scientific truth to the official ideology or sets itself up as judge of controversies between scientists, power succeeds merely in paralysing the progress of research conducted by its own subjects; a scientific suicide at the national level, which does not prevent truth from taking its course, research from progressing or its results from being exploited in other countries. In sum, if the state insists that the advancement of science must serve its own ends, it is thus acting against its own interests: German physics benefited no more from the distinction between 'Aryan science' and 'Jewish science' than Russian biology benefited from the contrast between 'bourgeois genetics' and 'Marxist genetics'. It is understandable that the partisans of 'laissez-faire' should point to the martyrology of science to assert that the less the state bears down on this field, the greater the chance of research developing; the slightest intervention already holds within itself the threat of overriding authority, of the transition from one scale of values to another, of the encroachment of the authority of force and bureaucracy into the authority of truth, the threat of objective alienation.[4]

But things are not as simple as that; the potential conflict between science and politics is not confined to a clash between the truth and its challengers. Truth is not in issue where there is only one truth: it is self-evident that science does not find its most favourable soil under a

3. Galileo, letter to Christina of Lorraine, Grand Duchess of Tuscany (1615), quoted in F. Russo, *Revue d'Histoire des Sciences*, 18, no. 4 (October–December 1964), P.U.F., Paris, 350 and 362.

4. See Michael Polanyi, *The Logic of Liberty*, Routledge and Kegan Paul, London (1951).

totalitarian régime. If the intervention of power were confined to this level, we should have no hesitation in concluding that science can develop only where the formal conditions of democracy are satisfied. And yet, if disciplines, theories, individuals and even teams may suffer—to the point of liquidation, as in the instance of Vavidov and his geneticist colleagues—from this interference with the affairs of science, other fields of research are not prevented from developing with as much success as under a liberal régime; the progress of physics in the USSR was not affected by the genetics crisis or even by the Stalinist purges of which some of the best physicists were victims. Conversely, the affairs of science are no less subject under liberal régimes than under totalitarian régimes to decisions imposed from without, and to interventions—more or less authoritarian, according to the nature of the régime, affecting the allocation of resources, the priorities assigned to certain sectors and therefore the general orientation of the research effort, but nevertheless unquestionable interventions, which cannot leave science indifferent.

Nowhere can political power dictate to science its procedures, the laws of its activity or its substance; power cannot determine the form or content of the scientific approach. The road to truth escapes political decision as much as truth itself; at a deeper level, truth—whether it be scientific or not, but especially if it is scientific—has its own authority which the authority of power cannot gainsay. Power may possibly hold it in check (by constraint or persuasion) by denying it access to public discussion, by screening part of its results or distorting its meaning, but no constraint or persuasion can change what it has established, save the authority of the scientific process itself.

> What Mercier de la Rivière once remarked about mathematical truth applies to all kinds of truth: 'Euclid is a real despot and the geometrical truths he has transmitted to us are really despotic laws'.[5]

The most tyrannous of princes will always be stopped short by that higher despotism. And yet, if political power cannot tell researchers *how* to search and still less what they can *find*, it everywhere aspires to tell them *what* they should look for—if only by guiding scientific personnel towards one field or discipline rather than another. Nowadays the despotism of truth is less frequently challenged than the fields in which it is called upon to exercise its absolute jurisdiction; the specific

5. Hannah Arendt, 'Truth and Politics', in *Between Past and Future*, The Viking Press, New York (1968), 240. (The quotation from Mercier de la Rivière is in French in the original.)

characteristic of the new relation which has become established between
science and politics is that *their conflict takes place no longer on the
ground of truth but also on the ground of performance.*

It will be retorted that this is not the ground on which pure research
should be required to render account of itself. But the question is precisely how one possibly could isolate the type of scientific research
which can be conducted without any solicitation or pressure from
society. The answer would be easy if there were any precise definition
of the different forms of scientific research. In theory, the more it is
directed towards practical objectives, the less reason it has to resist
social demands, and the less it aims at applications, the freer it should
be from external pressures. But scientific research is no longer made up
of distinct fields, separated by impassable frontiers and by gifts and
functions of differing character; on the contrary, these frontiers have
become so fluid that we no longer know where pure research ends or
where applied research, and even development, begins. If pure science
does not escape the criteria of utility which make research into an
institution linked with the production system, the reason is not only
that it finds itself assigned values which are foreign to its ends, but
also that *its own values are becoming more and more dissolved in the
functions it fulfils in achieved techniques.* The whole change in research activities since the second world war has challenged the idea of
pure science as a field of its own, different in essence from other research activities, *imperium in imperio.*

The place of pure research

If every definition of science refers back to the social organisation in
which it takes its place as end and means, it is perhaps surprising that
distinctions should still be drawn whose source and overtones are found
precisely in the options which modern science has tended to deprive of
all content. Just as science is deemed to be essentially intellectual, so
technology is deemed to be essentially physical: the tool is 'the continuation, the external projection of the limb',[6] and its intelligibility
seems to be wholly biological, whereas knowledge opens the windows
on another world, a realm detached from the biological constraints of
our own—pure intelligibility. In spite of the changes introduced into
the social function of science since the seventeenth century by the scientific revolution, the stereotypes inherited from antiquity persist; the
different forms of scientific research lie at different levels, 'pure' re-

6. Albert Espinas, *Les Origines de la technologie*, Alcan, Paris (1897), 46.

search at the top, technological research at the bottom. If there is a 'pure' science, there must logically be an 'impure' science, and all the words which may have been adopted to disguise this moral overtone merely succeed in shifting on to a different plane the ancient model of the intellectual city stratified according to a hierarchy of essences. 'Fundamental' science seems to be contrasted with 'trivial' science, 'basic' research with 'elementary' research, as though the drawing together of theory and practice had not affected the social values based on the assertion of an abyss between knowledge and action.

The hierarchy of mind and body, of idea and organ, still underlies the hierarchy of the savant and the engineer, even if the engineer shares in the researches of the savant or behaves like a savant in his own work; and particularly, even if the cultural concept of the *savant* is constantly merging more and more into the social function of the *scientist*. The new priesthood of researchers belongs to the same church, speaks the same language, is associated in the same enterprises, but it is no less hierarchical in its own image or in the image it presents to the lay world than it was before the upsurge of modern technology: from archbishop to country curate, the procession of researchers has not departed from the order of essences in becoming secularised.

And yet, pure science is merely one element among others in the *system* constituted by research activities: it no longer takes precedence on the road leading to the resolved enigmas of the universe. All contemporary research consists of reciprocal feedback between concept and application, between theory and practice, or, in Bachelard's words, between 'the mind which works' and 'the matter which is worked'.[7] In this relation *theoria* is the first instance of *techne*, in time if not in the hierarchical sense, and without its epistemological priority bearing a constant relation to the technical achievements which justify it; the road to the conquests of science lies through the conquests of technology. The experience of the war, and more recently space research or the big industrial laboratories such as Bell Laboratories, General Electric, Du Pont and IBM, has shown that while technical development depends closely on pure science, the progress of science depends just as closely on technology. The massive use of instruments has become the rule for scientists, just as daily recourse to concepts and theories has become the rule for engineers. It can no doubt be said that there is a greater degree of generalisation in pure science, but the development of certain tech-

7. G. Bachelard, *L'Activité rationaliste de la Physique contemporaine*, P.U.F., Paris (1965), 3.

nologies (nuclear energy, electronics, data processing) is constantly more dependent on wider conceptual equipment.

The tools of pure science are the source of technological assemblies which end by introducing into the economic circuit a new series of goods and merchandise either as the instruments of teaching or as consumer products (accelerators, telescopes, lasers, solar cells, computers); reciprocally, purely technical instruments condition the progress of theory at its most abstract levels. Just as science creates new technical entities, technology creates new families of scientific objects. The frontier is so fine that it is even impossible to distinguish between the mental attitude of the scientist and that of the engineer, so numerous are the intermediate cases. The scientist himself becomes an engineer, passing from pure theory to applied sciences and technology, as witness John von Neumann, Norbert Wiener, Claude Shannon and so many others; that is to say, he adopts in his own preoccupation with pure research the attitude of the engineer, just as the engineer must adopt the attitude of the scientist in his own research work. There is more to it than this; the rapprochement between theory and practice has changed the scientist himself into an entrepreneur in the industrial sense of the word, who has no hesitation in facing market competition on the basis of the discoveries he has made under the wing of the university, as shown by the vogue (and the financial success) of the science-based industries whose laboratories have multiplied near Boston, along Route 128 or in California. The mercantilism of the inventor–businessman, of whom Edison was the most complete example, has not spared the apparently disinterested realm of pure science.[8]

In antiquity, the distinction between theory and practice was grounded in a metaphysics; nowadays it is merely grounded in psycho-sociology. But if the motives and objectives of individuals do in fact trace out a frontier, it has no meaning except for researchers themselves; it is a purely subjective dividing line which society cannot judge and which is of no help whatever to the state in formulating a policy. Pure research, it is said, aims at the extension of knowledge for its own sake; if research is motivated by the desire for practical utilisation of its results, it becomes applied research. As Charles Kidd has shown, we thus arrive at two sorts of definition, each as vague as the other, which by

8. See Daniel Shimshoni, *Scientific Entrepreneurship*, stencilled doctorate thesis, Graduate School of Public Administration, Harvard University (1966), and 'La Route 128' (anonymous), in *Le Progrès scientifique*, D.G.R.S.T., Paris, no. 134 (October 1969), 10–52.

this very fact are quite meaningless to the authorities responsible for supporting a research project in proportion to its relation to applied research. The first definition turns on the researcher, the second on the content of the research; the criterion of the second is the more or less applied character of the product of research, that of the first is the personal motives of the researcher. It is vain for Kidd to show that definitions centred on the substance of research assume a confirmation of its nature after the event; we can decide upon it only if we know the end product and if we know that, within an indeterminate time, it has given rise to applications. Definitions centred on the motives and objectives of the researcher, for their part, lead to the inclusion in pure research of work which has nothing fundamental about it and which may be no less routine or trivial than some applied research.[9] To this it may be added that some applied research today leads to results which in the past would have been expected only from 'pure' research.

The situation is so confused that it has even been necessary to distinguish two categories within fundamental research itself:

> pure research, directed towards fuller understanding of nature and discovery of new fields of investigation, with no practical purpose in mind, and 'oriented' fundamental research, either focused on a given theme or background research directed towards increased accuracy of scientific knowledge in a particular field by gathering essential data, observations and measurements.[10]

A similar version put forward by Victor F. Weisskopf has the advantage of combining the criteria of the researcher's motives and the content of the research. He distinguishes two groups of basic sciences, which he calls 'obviously applicable' and 'frontier sciences'. The first group includes molecular biology, solid state physics, plasma physics and so on. These disciplines function as pure sciences, pursued with the aim of discovering basic phenomena and without any practical applications in mind, though it is obvious that any progress will have a practical effect (molecular biology in medicine, solid state physics for the production of new metals and plasma physics in exploiting the energy of nuclear fusion).

The second group, 'frontier sciences', includes high energy physics, galactic and extra-galactic astronomy (the astronomy of the solar system

9. C. V. Kidd, 'Basic Research—Description versus Definition', in *Science and Society* (ed. Norman Kaplan), 146–55.

10. Pierre Auger, *Current Trends in Scientific Knowledge*, Unesco, Paris (1961), 245.

being today 'obviously applicable' to space research), cosmology and so forth; these sciences deal with objects so distant from the phenomena that occur on earth that they have no obvious applications. Such applications cannot be ruled out; the knowledge apparently least relevant to practical utility is never without possibilities of practical application. Recalling Rutherford's remark in 1933 that no source of power could be expected from the research in atomic physics to which he himself had opened the way, Weisskopf points out that the distance in space of the objects of the 'frontier sciences' no doubt involves a distance in time in the utilisation of their possible results, but it cannot be asserted *a priori* that these sciences have no practical effect.

> Nobody knows what will happen when it becomes possible to produce beams of particles thousands of times more intense than those of today— beams in which all the new particles will be mass-produced to react with the environment. Even now some people are speculating about the special clinical effects of pion beams, effects that cannot be produced by ordinary radiation.[11]

The only distinction which still seems defensible is that based on the different channels of information and communication appropriate to science and technology. The channel of science is in principle open, based on criticism by peers, easily and rapidly accessible to all researchers in the specialised literature, while the channel of technology is less available, more subject to the restraints of industrial organisation and competition, bound by secrecy or, more simply, by the difficulty of transmitting the subtleties, the 'know-how' of processes that depend more on apprenticeship on the job than on the understanding of concepts. It has even been said that 'the scientist wants to write and not to read and the technician wants to read and not to write',[12] thus emphasizing what discovery owes to its publication, the scientist being concerned to reveal what he is doing, while the engineer refrains and is guided by the recipes published in the technical papers, or even advertising leaflets, rather than by scientific literature.

But this definition does not take any greater account of intermediate cases; the publication of a discovery may be withheld by imperatives

11. V. F. Weisskopf, 'Is Pure Science doomed to decline in the U.S.?', *Scientific American*, **218**, no. 3 (March 1968), 139–44.
12. Derek J. de Solla Price, *The Difference between Science and Technology*, Thomas Alva Edison Foundation, Detroit (1968), and 'Research on Research,' in *Journeys in Science: Small Steps—Great Strides* (ed. David L. Arm), University of New Mexico Press (1967), 10–11. See also Harvey Brooks, 'Applied Research: Definitions, Concepts, Themes', in *Applied Science and Technological Progress*, National Academy of Sciences, Washington (1967), 36–40.

of secrecy imposed for reasons of either national or industrial policy, while technological innovations may be the subject of free and rapid publication in the scientific press. The difference in channels of information and communication perhaps defines a different scientific culture, but it does not rule out similar research practice. It is clear that the evolution of technological research, on the contrary, is tending to reduce if not to suppress this opposition. In fact there is a constant and inescapable cross-traffic between the two fields, the import and use of concepts in technology, the transfer and handling of instruments in science—and their transfer from one science to another: radioactive tracers have transformed chemistry and biology, just as the discovery of new elements has transformed the technology of materials.

In a sense, Pasteur foreshadowed the difficulties of these definitions when he protested against the converse excess of reducing the edifice of science to the applied sciences only:

> No, a thousand times no, there is no category of sciences which can be called applied sciences. *There is science and the applications of science, bound together like the fruit and the tree which bears it.*[13]

But the tree of science also hides the forest: scientific research does not consist solely in science in the sense of knowlege which is the sole source of applications. *From the most abstract reflection through to development, scientific research constitutes a process whose different elements are so many links in a continuous and retroactive system.*

> The two societies, the theoretical society and the technical society [says Gaston Bachelard, with great truth], are in contact with each other. They cooperate. They *understand* each other.[14]

This does not, of course, mean that they *recognise* each other; in the eyes of some researchers, as in the public image, it is always the savant, the scientist, who enjoys greater prestige than the engineer. This comprehension is not *natural*, Bachelard goes on to say, thinking first of the intellectual horizon in which contemporary physics is defined as anti-nature, artifice, the 'fabrication of phenomena'. But in his mind this horizon could not be separated from the social environment:

> Rational objectivity, technical objectivity, social objectivity, are henceforth three closely linked characteristics. If one disregards any single one of these characteristics of scientific culture, one immediately enters the realm of utopia.[15]

13. *Oeuvres de Pasteur* (ed. P. Vallery-Radot), vol. 7, 215(Pasteur's italics).
14. G. Bachelard, *L'Activité rationaliste de la Physique contemparaine*, 9.
15. *Ibid.*, 10.

It must be added that social objectivity is steeped in political definitions, that it is defined in the inevitable space of state intervention and
orientation by the state. To forget this is to stray into the realm of
mythology. In the eyes of the scientist, pure research constitutes a self-
contained world whose motives, objectives and norms have nothing to do
with other forms of scientific research, although this world is no less
bound up with 'social objectivity' than the world of applied research.
Just as knowledge is inseparable from the techniques which are
grounded upon it and by which it is nourished, so it cannot be dissociated from the social context in which it develops; the whole reward
for the success of pure research has been its integration in the general
research system. In the eyes of power, the function of research is not to
seek but to find; it is idle for pure research to visualise itself as an ideal
intellectual activity, the source of truth, when in fact it fulfils itself as a
practical activity, the source of applications.

Many scientists, says Don K. Price,

> tend to look back on pre-war science as the Reformers looked back to the
> Primitive Church: a period of austere purity, an era in which no vows
> were needed to guarantee the poverty of the professor, no scientist was
> seduced by a government contract, and teaching fellows were obedient.
> One may well be a little sceptical about this point of view, and suspect that
> poverty probably brought its distractions no less troublesome than those
> of riches.[16]

The longing for the 'good old days' of research may well stem from a
conception of pure research which distinguishes it from applied research
and above all from technologies, and which asserts its independence of
the needs, aspirations or demands of society, but the fact nevertheless
remains that it is increasingly difficult to divorce the future of science
from the future of technology; the way in which their interests inter-
lock, even if scientists think and act as though they were conflicting,
induces the state to associate them in a common destiny.

Free research and oriented research

Scientific research can best be characterised in relation to other human
activities by combining themes specific to information theory and economic theory; it is work directed towards an uncertain product. Strictly
speaking, this activity has no material product, and its output consists
solely of information: ideas, theories, models, plans, reports, documents, papers, that is to say, products whose space–time pattern is not

16. Don K. Price, 'The Scientific Establishment', in *Scientists and National
Policy-Makers*, Columbia University Press (1964), 39.

that of accountable goods. In the eyes of economists, the main distinction between knowledge and material goods is that knowledge does not wear out; you can use it as often as you like and it can be replaced only by fresh knowledge which makes it obsolete. This obsolescence, moreover, is never total, since all new knowledge merges with the knowledge it renders obsolete rather than replacing it; the product of research is never completely wiped out by a new product. Furthermore, while material goods are confined to one place, and can be used only one at a time, research results are never exclusively appropriated; knowledge is omnipresent. In this sense, the work of scientific research has an intellectual history which social determinations fail to take into account; there is a whole genealogy of concepts, methods and scientific theories which is specific to the internal history of science. But not only does scientific research have its own intellectual history, it also produces concrete goods; the flow of information spills over into social practice, and new knowledge—discovery and invention—ends in action.

There is no longer any break in the process from the extension of pure knowledge to the creation of new families of technical objects. It is as hard to distinguish between discovery and invention as it is to distinguish between pure research and applied research. Is the dividing line the degree of generality and coherence? But in many cases invention is no less dependent than discovery on theoretical generalisation. There are still a great many inventions which owe little to science, but there are more and more which owe it a great deal. What Bachelard said about a rationalised theory of electricity in fact applies to the whole structure of contemporary scientific research:

> The rational and the real must be apprehended together in a veritable coupling in the electromagnetic sense of the word, constantly stressing the reciprocal reactions of rational thought and technical thought.[17]

This idea of coupling represents movement from the most theoretical to the most applied research and movement back again not as a fortuitous transition from intellectual adventure to technology, but as *the deliberate organisation of reciprocal exchange*. Technical innovation results precisely from these constantly reciprocal relations between the rational idea and its industrial application, a continuous process which has neither beginning nor end, and develops through a series of events each of which answers the questions raised by previous efforts while at the same time raising new questions of its own (the end-product of re-

17. G. Bachelard, *Le Rationalisme appliqué*, P.U.F., Paris (1966), 139.

S.P.—4*

search is the subject of constant technical improvements). Scientific research results in discoveries which are also inventions, in inventions which are also discoveries: *it is the deliberate and organised application of human labour to the production of new knowledge, processes and products.*

The social translation of this evolution is the concept of 'Research and Development' ('R & D') which lumps together under the same label all research activities starting with the investments they need. Science policies—state responsibility for, and orientation of, scientific research—are expressed in those statistics in which all research activities, however remote from practical application, are judged by the same criteria of utility. The political approach which thinks of the research system as an integrated whole, from pure research to development, by way of applied research, postulates not only that knowledge is directed towards action, but also that there is a relation of cause and effect between the support it enjoys and its capacity to adapt itself to practical application. From the standpoint of utility, under which the quest for new knowledge assumes the character of a joint venture, there can be no criterion except the return on investments.

In considering pure research, whose results and their effects are unforeseeable, there is, however, no way of measuring this return in either the short or the long term. It makes no difference; the hope is still cherished of arriving at some measuring rod, just as though the process of discovery could be mastered by the mere fact that its results have become generally exploited. 'Project Hindsight', carried out by the United States Department of Defense, illustrates to the point of absurdity the limitations of this approach. This 'flashback' survey—hence its name—was designed to identify the contributions of the different agencies (government, industry, universities) and the different types of research to the finalisation of the new weapons systems developed over twenty years.[18] Fifteen systems have so far been analysed to identify the 'events' or scientific contributions which have reduced their cost or helped their development. Of these 'events' 92 per cent have been identified as technological and 8 per cent as scientific. In this last category the great majority of the 'events' arose out of applied research directly connected with the needs of the Department of Defense (6.2

18. C. W. Sherwin *et al.*, *First Interim Report on Project Hindsight*, Office of the Director of Defense Research and Engineering, Clearing House for Federal Scientific and Technical Information, Washington (June 1966), and Raymond S. Isenson, *Final Report*, National Technical Information Service, Department of Commerce, Washington (October 1969).

per cent), another group out of applied research for commercial purposes (1.5 per cent), and only two events (0.3 per cent of the scientific
contribution) could be associated with pure 'non-oriented' research. It
is easy to see why the preliminary report on this survey immediately
spread consternation among the United States scientific community: it
tended to prove that the enormous sums lavished by the Defense Department on pure research had relatively little effect on the perfecting
of advanced weapons systems, whereas investment in 'oriented' research
had paid off more than handsomely. From there it was only one step to
the conclusion that money spent on pure research for the purposes of
military programmes was money thrown away.

And yet this survey did not prove very much from the point of
view of pure research, except perhaps that its fate is not necessarily
bound up with the military budget or at least that military programmes
do not necessarily benefit from pure research. In the first place, the
specifications for these programmes, by definition, call for recourse to
essentially technological 'events'; the atom bomb did not spring fully-
armed from Einstein's equation, like Athene from the forehead of Zeus.
Furthermore, the criterion of return on a process or product in the context of defence can be applied to a discovery only at the cost of con-
founding its scientific value with the value of the applications to which
it has led. A weapons system (rockets, torpedoes, mines, bombers,
radar) may become x times cheaper and y times more effective thanks
to the progress of scientific and technical research, but neither the
reduction in cost nor the increase in effectiveness resulting from a discovery is an adequate measure of its advantages as a contribution to
the stock of common knowledge. Finally, and above all, it is completely
arbitrary to fix some given date (in this case 1945) as the starting point
for the contribution of science to recent weapons systems when it is
perfectly obvious that none of the 'scientific events' and not all of the
'technological events'

> could have occurred without the use of one or more of the great system-
> atic theories—classical mechanics, thermodynamics, electricity and mag-
> netism, relativity and quantum mechanics.[19]

Along these lines, as the authors of the survey somewhat ingenuously
admit, one might as well go back to Newton's laws, Maxwell's equa-
tions or Ohm's law, 'events' which would have come into the picture so
frequently that they would have completely masked all more recent

19. C. W. Sherwin and R. S. Isenson, 'Project Hindsight—A Defense Department study of the utility of research', *Science*, 156 (23 June 1967), 1576.

'events'—not to speak of less outstanding results which, even if they are not ranked as 'events', have none the less helped to increase the stock of knowledge on which the new weapons systems could draw.

This example is significant not so much for what it sets out to prove as for what it unwittingly reveals; the criterion of utility by which Research and Development is defined as a subsystem of the production system compels us to look for a mathematical relation between investments and scientific results, but this relation can never be established, even very roughly. A calculation of this kind is already very difficult for applied research in an industrial firm and much more so at the level of a country. The idea that a calculation of this kind can be made for pure research is, in Bergson's language, nothing but treating in quantitative terms what is a matter of duration. No doubt the relative production of scientific knowledge can be measured as between one country or group of countries and another, on the basis, for example, of the number of articles published in specialised reviews, and it is even possible, from these statistics, to form an idea of the quality of the scientific work measured in this way.[20] But it is quite another speculation to profess to establish a sort of 'cost-benefit analysis' relation between pure research and the application of its results.

The authors of Project Hindsight suspect that the primary impact of science

> may be brought to bear not so much through the recent random scraps of new knowledge, as it is through the organised, 'packed-down', thoroughly understood and carefully taught, *old* science.

They are none the less determined to bend the search for knowledge to a system of evaluation which determines its present performance:

> When one debates the utility of science, the real issue is not the value, but rather the time to utilisation.[21]

Thus, the idea, the theory, the discovery, are no longer judged except by the time it takes to apply them in practice; but if utility is consecrated only by conversion into act, what is it that makes that act a useful act? Since pure research is a matter of qualitative time rather than of quantitative time, its product has a duration which does not come to an

20. See Derek J. de Solla Price, *Little Science, Big Science*, Columbia University Press, New York (1963), and Joseph Ben-David, *Fundamental Research and the Universities*, OECD, Paris (1968).

21. 'Project Hindsight', *Science, loc. cit.*, 1576. See also Karl Kreilkamp, 'Hindsight and the Real World of Science Policy', *Science Studies*, 1, no. 1 (January 1971), which is the best evaluation I know of this work.

end with its first applications. In the first place, this product is, by definition, haphazard. Secondly, since knowledge is omnipresent, new knowledge has many uses in an indeterminate period of time; the results are not necessarily exploited in practice in the same country in which they were first arrived at, and these results themselves open the way to applications which were not necessarily foreseen when they were first exploited.

Summing up, what Project Hindsight best proves is that, from the point of view of sociological methods, the results of a survey tend to conform to expectations and therefore to the interests of the institution which sponsors it! *Traces*, a similar survey, but conducted for the National Science Foundation whose primary function is to subsidise non-oriented research, led to completely opposite results. Whereas Project Hindsight dealt with military technology, disregarding 'events' before 1940, *Traces* concentrated on five 'economically and socially important' non-military innovations (magnetic ferrites, video tape recorder, oral contraceptive pill, electron microscope and matrix isolation) without setting any forward date limit to their genealogy.[22] Out of 341 identified 'events', *Traces* concludes that 70 per cent were the result of 'non-mission' research (compared with 0.3 per cent for Hindsight!), 20 per cent of 'mission-oriented research', and 10 per cent of 'development and application'. Now, since 76 per cent of non-mission research was conducted in the universities, in the aggregate more than half of all the scientific and technological 'events' which led up to these five innovations originated in the universities (compared with 9 per cent for Hindsight!). *Traces* further concludes that more than 90 per cent of all 'events' connected with non-mission research were completed before the finalisation of the innovations, while 10 per cent were completed ten years after they first appeared.

This goes to confirm that the process of technical innovation largely depends, not only for its origin, but also for its progress, on basic research, that is to say, on a stock of knowledge and theoretical work whose prior existence does not prevent it from continuing to condition the appearance and development of new technologies. *Traces* thus suggests that the 'educational cycle' of 'events' connected with basic research is some thirty years, these 'events' reaching a maximum between the twentieth and the thirtieth year before the innovation, while the

22. 'TRACES'—*Technology in Retrospect And Critical Events in Science*, prepared for the National Science Foundation by the Illinois Institute of Technology Research Institute, NSF–C535, Washington (December 1968).

last ten years are richest in 'events' connected with applied research which determine the finalisation phase of the innovation: 'Most inventors rely heavily on information created in the previous generation'.[23] The orientation of the two surveys is so different that the divergence of their conclusions is hardly surprising; Hindsight plots the history of an innovation within the limits of its development, *Traces* in the duration of its conception. But *Traces* was not concerned with either the financing conditions or the cost of the research activities under review, so that, while its methods may clearly appear to challenge, on solid grounds, those of Hindsight, its results do not in themselves prove that the Hindsight results are open to challenge on the ways and means of influencing the 'educational cycle' to cut down the delay between conception and application.

In fact, the only conclusion which can be drawn from Project Hindsight is the relation it establishes between the time it takes for basic research to produce results and the assignment of the sponsoring agency; the more closely research (whether basic or not) is linked with the aims of the sponsoring agency, the more quickly it is likely to yield useful results. The great majority of the identified 'events' were in fact the product of financing by the Defense Department, and of research based on a clear understanding of the department's needs. In other words, the 'cost–benefit analysis' of pure research should be directed less to its results than to the needs satisfied by those results.

This conclusion to some extent coincides with that reached by the economist J. Schmookler; studying technological innovation in four branches (railways, agriculture, petroleum industries and paper), Schmookler shows that it has been introduced in response to economic demand rather than as a direct result of basic research.[24] Science does not automatically lead to practical consequences from the mere fact that it develops under the promise of practical applications; *the time lag before it is used depends not so much on the importance attached to it as a source of new knowledge as on the needs which it can satisfy as a fully realised technique.* The example of nuclear research is no doubt the most telling: most of those who contributed to it had not the slightest idea, before uranium fission was demonstrated, that their work might lead to important applications; it nevertheless took the war effort to convert theoretical and experimental research into a rapidly

23. *Ibid.*, 12.
24. J. Schmookler, *Invention and Economic Growth*, Harvard University Press, Cambridge, Mass. (1966).

industrialised technological enterprise. Other examples could equally well be cited, from radar to computers; the chances of transition from scientific discovery to innovation depend more on demand external to knowledge than on its intrinsic virtues.

Among these demands, the threat of war and the challenge of strategic situations (polical rivalry or economic competition) have always been the most compelling. But they are not alone in acting as a stimulus to technological transfer. In fact the strategic model of urgent situations at national level ended by inspiring the programming of research activities at the level of industrial firms; the cost-effectiveness techniques applied to military programmes are being increasingly applied to civilian research programmes. Advanced societies are precisely those which organise themselves scientifically to cut down the interval of time between pure science and the application of its results; hence the great success of 'technological forecasting' which aims at predicting discoveries as much as the needs which may be satisfied by their possible applications.[25]

Supported with an eye to its practical results, pure research must bow to the criteria of performance which govern the whole production system; its theoretical proceedings are judged by the applications to which they lead, even if its scale of values has nothing to do with utility. It can be said that the pressure to which it is most commonly subject, whatever the political régime in which it develops, is the desire to speed up this transition from theory to practice. In liberal societies, as in totalitarian societies, the scientific process is now challenged less often on the grounds of truth than on the grounds of its pay-off. The problem of the freedom of science is no longer *solely* that of the clash between truth and its enemies, it is also the problem of the conditions in which pure research is allowed time to fulfil itself in the form of technology; the imperative of performance is the industrial avatar of the threats which the authority of dogma has always brought to bear on the scientific approach. *Relation to truth is therefore no longer the researcher's sole guarantee against the encroachment of power; he also needs an institutional framework which recognises his freedom in relation to time.*

In the last analysis, it is clearly the environment in which they are conducted, rather than their motives, goals or procedures, which dis-

25. See E. Jantsch, *Technological Forecasting in Perspective*, OECD, Paris (1967), and our discussion of the methods and ambitions of this kind of forecasting in chapter 5.

tinguishes between the different forms of scientific research.[26] The university is in principle the institution which allows the greatest degree of freedom to work aiming at no immediate or foreseeable application, but an industrial laboratory—and even a public laboratory—may shelter research teams whose activities are no more 'oriented': 'freedom of time' depends not on the nature of the institutions, but on the tasks which they set themselves at any given moment. Scientific research, as Harvey Brooks so rightly says, is a continuous process involving a series of contingent choices on the part of the researcher; every time he decides between alternative lines of action, the factors which influence his choice determine how far his activity will be basic or applied:

> If each choice is influenced almost entirely by the conceptual structure of the subject rather than by the ultimate utility of the results, then the research is generally said to be basic or fundamental, even though the general subject may relate to possible applications and may be funded with this in mind.[27]

The research which led up to the discovery of transistors in the Bell laboratories called for work which, in the university context, would have been classified as basic. In the eyes of the sponsoring firm, however, it was regarded as applied, purely because of the existence of potential customers. Brooks emphasises, moreover, that the detailed choice of the different stages of research would no doubt not have been the same in the two environments, and might have led to different lines of inquiry starting from the same point:

> If the next step is toward the particular, the research is more likely to be applied, but if it is toward the general, or toward widening the scope of applicability of a technique or principle, it is more likely to be basic.[28]

Thus, the research man in a semiconductor industry might concentrate on the purification of promising semiconducting materials, while the university scientist might become more interested in exploring a wide variety of materials. A recent example tends to narrow the distinction even further; the first enzyme synthesis was achieved simultaneously in 1969 in a university laboratory (Rockefeller University) and in an industrial laboratory (Merck). Applied in the one case and basic in the other, this research can no longer be distinguished except by the institution which houses it.

26. As an application of this theme, see J.-J. Salomon *et al.*, *The Research System—Comparative study of the organisation and financing of fundamental research*, vol. 1, *France, Germany, United Kingdom*, OECD, Paris (1972), and vol. 2, *Belgium, Netherlands, Norway, Sweden, Switzerland* (1973).

27. Brooks, *op. cit.*, note 12, 23.   28. *Ibid.*

As definite categories, basic and applied research tend to be meaningless, but as positions on a scale within a given environment, they probably do have some significance.[29]

There is no pure research *per se*, but research activities whose character is defined by the institution that sponsors them. In this sense, the research system includes only two types of activity: *free* research, in so far as the sponsoring institute assigns it no practical end (the practical end being often beyond the imagination either of the institute or of the researchers), and *oriented* research, whose applied character becomes more marked as the objectives set by the sponsors become more precise. There is no gulf between the two other than the difference between the aims of the sponsors; theoretical physics may be free or oriented according to whether it is researched in a university laboratory or an industrial laboratory, virology according to whether it is practised in the Institut Pasteur or in Hoffman La Roche.

But this free speech itself is subdivided into two very different categories according to the volume of resources it requires; the scale of investments traces a dividing line which is not based on any difference in the attitude either of the researchers or of the institutions that shelter them. A distinction is drawn between 'academic research', which requires only limited equipment and depends essentially on individual activities, and 'programmed basic research', organised around very costly equipment and large research teams, often multidisciplinary, and a great deal of technical help. This second category belongs to the era of 'Big Science'; it is not explicitly aimed at rapid application, any more than the first category, but it differs from it by the numbers of researchers and the amount of instruments and investments it requires. For example, the cost of installing and operating a large accelerator runs to several hundred million dollars; it takes nearly two years to draw up the plans, up to eight years to build it—and sometimes five years to carry out a single experiment in a bubble chamber and exploit its theoretical results. It is clear that the freedom of this research has an economic limit; it cannot be independent of political decisions. The construction of a big accelerator, or participation in its creation, depends on a decision which cannot take account solely of the wishes of researchers or the interests of science. The choice is so important that it can be made only at government level. A balance must be struck, first among the resources allocated to all research activities, and second, between research and all other fields of government expenditure. Free research may very well not be directed towards any specific end, but it is

29. *Ibid.*, 24.

none the less conditioned by the goals of power; 'there is no question here of planning or not, but of planning well or badly'.[30]

There is therefore only a very small fraction of the research system which does not, in theory, depend on decisions reached outside the 'research community', and it is only to this fraction that the idea of the autonomy of science would still apply if, in fact, the decisions affecting big science did not in turn also affect the equilibrium of little science. The authoritarian allocation of *resources* is also the authoritarian allocation of *values*; the options reflected in the budget of the research system have their repercussions, step by step, on all activities. With the possible exception of research which needs no equipment except a blackboard or pencil and paper, the whole research system depends on decisions which are not purely a matter of science and on objectives which are not those of researchers alone. From this point of view free research is no more able than oriented research to escape the criteria of social utility by which the state defines its goals. The services it renders are not accounted for in the form of material goods, but the investments it needs are accounted for in so far as they might be appropriated to other purposes, whether other sectors of research or different kinds of activity; free research is free only on these terms.

30. *Governments and the Allocation of Resources to Science*, OECD, Paris (1966), 21.

CHAPTER 5

Science Policy and its Myths: Planning and Forecasting

'The fate of our times', says Weber, 'is characterised by rationalisation and intellectualisation and, above all, by the "disenchantment of the world".'[1] The triumph of science as a technique of forecasting has been paid for by the breaking of a spell; no longer is there any comprehensive image of the universe whose meaning can be revealed to us by rational knowledge. Scientific activity is no royal road to the gods or to the essence of the world, or to the truth of nature hidden behind its outward appearances; the charm of mysterious powers waiting to be discovered or invoked has faded away with the multiplication and success of objective, neutral and purely technical means of mastering natural phenomena. If science robs the world of enchantment, it is because the only answers it offers to the questions we ask of it are instrumental ones.

But science itself is caught up in this process of disenchantment; as one technique among many others, it cannot escape the effects of its own action in rationalising nature as quantifiable and predictable matter, submissive to control and organisation. Just as scientific activity is designed to reduce uncertainty, so it must itself submit to the mathematical apparatus designed to reduce the uncertainty of research or to introduce some order among its different directions. The time has gone when Benjamin Franklin could retort to the sceptical bystanders who asked what was the use of an early balloon flight by retorting 'What is the use of a new-born child?' Science no longer has to find justification for research, it has given proof of its utility—it has delivered the goods. But perhaps it has proved too much: the new-born child is valued only for its potential adult productivity, and the care lavished on it is motivated only by the hope of a quick return.

If the political authorities are interested in the sources and genesis

1. See Max Weber, *Science as a Vocation*, 155.

of knowledge, it is only because of the uses to which that knowledge can be applied; in spending money to increase the supplies of skilled manpower, facilities and research potential, they are thinking not so much of the advancement of knowledge for its own sake as of the goods through which that advancement is translated into new or more powerful instruments of action. What they have in mind is the last stage of the research process, when the discovery or invention is exploited and becomes an innovation, and is given reality in the form of products or processes which are added to the arsenal of means of production or of destruction. Scientific activity is fitted into the same pattern of thought which thinks it possible to express the aims of society in economic language in terms of the resources available to achieve them; efficiency, productivity, maximisation are the frame of reference by which the research system must establish its social justification. When the resources allocated to research and development reach two per cent or more of the gross national product (in the United States and the USSR, more than three per cent), the individual adventure of research quite obviously involves collective decisions; the scientist is no longer answerable only to his peers or to scientific truth, he must plead before the tribunal of the community, which determines budget appropriations and controls the use to which they are put. Thus, the research system itself is no more than a subsystem of the network of economic relations over which the state exerts its authority. Even if the 'goods and services' which it provides do not fit neatly into the classical definitions of national accounts, the money spent on it is clearly an item in the public accounts.[2]

It is intellectually tempting—it is the whole temptation of rationalism—to imagine that the approach of the public authorities towards science is itself scientific, that it is science which determines the ways and means by which support is given to knowledge whose advancement depends more and more closely on the state, that it is rational technical action which fixes the procedure and sets the goals in the light of which the adventure of discovery is embodied in a collective institution. If insistence on rational management is one of the characteristics of

2. 'The national accounts only show goods and services which are effectively exchanged or capable of being exchanged on the market.' (*Comptes de la Nation*, Paris, Ministère des Finances (1960), vol. 2, *Méthodes*, 150.) The United States definition is not very different: 'The fundamental criterion used to define an activity as economic production is that which is reflected in the transactions of buying and selling on the market.' (*National Income: Supplement (1954) to the Survey of Current Business*, Washington, 30.)

modern society, this insistence will seem all the more legitimate when it relates to a social system whose members, values and institutions are by definition dedicated to reason.

And yet, science policy, in thus confusing the technical representation of knowledge with the human scene of power, is merely creating myths; political processes do not automatically become more rational by the mere fact of being applied to the scientific world. Myths or the delusions of the technocratic ideology of management? Some of the themes which recur most frequently in discussions on science policy—planning, forecasting, the criteria of choice—might create the impression that the will to rationalise which is peculiar to the government of modern societies depends for its fulfilment only on the perfection of knowledge and techniques. It is not that, as Marcuse would understand it, the triumph of positive thinking in this case turns the rational into a support for the irrational; at this level, there is no turning of rationality into its opposite [3] but rather the demand for an accomplished rationality, accessible or operational in its very accomplishment—in other words, scientism in its most traditional sense. The limits and deficiencies of decisions are attributable to defects in our information and instruments of action, not to the nature of things, of men or of society. If we cannot do more today, it is because we do not know enough; tomorrow, management techniques will have progressed to such a point that we shall be able to master the most complex systems and storm the last redoubt of the unforeseeable. . . . It is, however, enough to review some of these themes and the way in which they have aroused the hope of a 'political technology' of science to realise how far the research system, even regarded as a productive force among many others, still resists any effort to rationalise it.

Planning and Programming

It is idle to ask whether science can be planned or not; the question is purely rhetorical, as Harvey Brooks said, since

> science *is* planned, whether implicitly and largely as a by-product of decision-making processes external to science, or explicitly and consciously.[4]

From the moment that the state steps in as the main supporter of research, the allocation of appropriations among disciplines, projects and

3. Herbert Marcuse, *One-Dimensional Man*, Beacon Press, Boston, Mass. (1958), and Routledge, London (1964), especially chapter 7.
4. Harvey Brooks, 'Can Science be Planned?', *Problems of Science Policy*, OECD, Paris (1964), 97.

researchers is not determined solely by scientific skills. Theoretically science planning aims at

> the best adjustment between the need of science for internal autonomy and the desire of society for the fruits of science.[5]

But in the first place scientific research is only one dimension of planning, and in the second place the aspirations of the research system must take account of the aims pursued by the state, that is to say, of criteria which relate not to science itself but to its applications.

In the best of all possible worlds, even plenty would not do away with the problem of choice. Between the competing demands of different sectors of activity, public support is determined not only by the amount of resources available but also by the orientations freely chosen or imposed by the economic situation. The dilemma of Buridan's ass is no problem for governments; decisions are taken one way or another, even if it is only the decision not to decide. The aims of the state should as far as possible be mutually consistent, but they are no more homogeneous than the needs they are designed to meet. Defence, transport, social security, education, scientific research, and so forth, are all so many heterogeneous and incommensurable demands, whose pressure on the public budget is uneven, varying according to circumstances and the short-term targets set. Furthermore, within each of these sectors there is an open option between alternatives which are no less heterogeneous and incommensurable. It is impossible to do everything at once or for the same reasons, since at any given moment the resources are always less than the needs which emerge. What passes for coherence in a policy is never more than a precarious balance between widely differing needs.

> The characteristic attitude in large-scale economic management, both inside government and in the private sector, which has made itself increasingly felt during the post-war period, is the pursuit of intellectual coherence. Its most obvious manifestation is in long-range national planning.[6]

Whatever the differences in planning institutions, techniques and styles in different countries, planning everywhere is designed 'to reduce the area of the unpredictable to a manageable series of clear alternatives'.[7] A 'calculated risk, a fail-safe device', in the words of Pierre Massé ('*aventure calculée, anti-hasard*'),[8] planning holds itself out in every case, whether it be imperative or indicative, flexible or inflexible, as the

5. *Ibid.*, 97.
6. Andrew Shonfield, *Modern Capitalism*, Oxford University Press (1965), 67.
7. *Ibid.*
8. Pierre Massé, *Le Plan ou l'Anti-Hasard*, Gallimard, Paris (1965).

instrument of rationally formulated choices. And yet, the rationality of planning experiments is never established—if at all—until after the event.

At the level of a business firm, it is relatively easy to query the object of the costs to be incurred, to set priorities among targets and to acquire the means of achieving them; the cost of the various methods envisaged and the estimate of potential return are circumscribed within the microeconomic area of the exclusive interests of the firm. Similar methods may even be applied to the choices which must be made by a specialised ministry, as witnessed by the success of the 'functional' budget techniques introduced by the United States Defense Department. These methods make it perfectly feasible to compare the relative cost and benefits of different programmes designed to achieve the same end.[9] But as soon as one tries to generalise this management technique by comparing the advantages of programmes designed for different ends, the choice comes up against the limits inherent in all decisions relating to the macroeconomic area. Not only does the calculation lose strictness by introducing a constantly growing number of variables but, above all, it meets an insuperable limit; you cannot measure the incommensurate.

A public authority, such as an electricity board, may try to analyse with mathematical precision the costs and benefits of a specific type of power station, thermal, hydroelectric or nuclear, or a railway operator may analyse the costs and benefits of various types of line, even calculating their marginal return and the financial incidence of variations in fares and freight rates as a result of road competition. So long as we remain in this way within the limits of allied activities, connected with a common purpose, the search for coherence is all the more likely to succeed, since it is limited in practice to defining the quantitative implications which it is the whole object of planning to maximise. It is not surprising that these budgeting techniques should have developed as a consequence of strategic arms programmes, where it is possible for a given objective to translate into formulæ and mathematical models the means which will provide the desired product in the fastest and cheapest way. In such a context, say the theorists of *The Economics of Defense in the Nuclear Age,*

9. See in particular David Novick, *Program Budgeting: Program Analysis and the Federal Budget,* Harvard University Press, Boston (1965). A form of the Planning–Programming–Budgeting System is now being applied in France under the name of *Rationalisation des choix budgétaires (RCB).*

the choices that maximise the attainment of an objective for a given budget
are the same choices that minimise the cost of attaining that objective.[10]

In other words, there is no incompatibility between the economic,
technological and strategic considerations which enter into the finalisation of a weapons system, 'the strategy which is most efficient being
the most economical'.[11] The problem of security being defined as an economic problem among others, the search for coherence is so completely merged with the desire for efficiency that efficiency ends by
taking the place of coherence.

On the other hand, when it comes to comparing different activities
whose purposes do not appear to converge in any way, the chances of
achieving this coherence are already lessened by the greater number of
variables to be dealt with, just as games theory finds itself doomed to
greater uncertainty with the introduction of new players. But in addition, and above all, the lack of uniformity in the terms fed into the
comparison rules out the presentation as a rational calculation of what
will always remain subjective and ideological, namely political choice.
What is there, for example, in common between the construction of a
power station or a railway line, and that of a school or hospital, between
support for wheat producers or growth industries and the budget for
cultural activities or military programmes? One does not have to be an
expert in formal mathematics to know that the problem of coherence is
not that of the system in which we take our stand, but of the links between different systems. It is idle to profess that the search for economic
coherence expresses a social rationality, which is something that cannot
be translated into fact by mathematical calculation, no matter how objective or rigorous it may be; the incompatibility between the rival needs
and aspirations of the different systems cannot be cured solely by the
criteria of economic efficiency.

It is juggling with words to talk of 'planning' where, in reality, there is
still only 'programming', and *a fortiori*, to talk of rationality when it is
only a question of eliminating those public expenditures which are less
justified or more costly than others. In this sense, 'rationalising' the
national budget is merely a management problem. However perfected
they may be, management techniques are never enough to transfer the
rationality of the means adopted to the ends which they are supposed

10. Charles J. Hitch and Roland N. McKean, *The Economics of Defense in
the Nuclear Age*, Harvard University Press (1960); reissued by Atheneum,
New York (1965), 2.
11. *Ibid.*, 3.

to ensure. There is no conflict of interest, say the authors of *The Economics of Defense in the Nuclear Age*, between budget considerations of economy and military considerations of effectiveness, 'except in the determination of the *size* of the budget or the magnitude of the objective to be achieved'.[12] But is not this the whole problem of the coherence at which planning aims? Inside a given system (defence, education, research, etc.) planning takes the form of programming the resources and means which will allow a given objective to be achieved most effectively, rather than of setting those objectives. As soon as a number of systems are confronted with one another, the choice of a 'manageable series of clear alternatives' to which planning aspires gives rise to arbitration between conflicting interests in conditions of imperfect information, ambiguity, negotiation and divergent pressures which inevitably characterise the political fabric of the process of allocation resources.

Just as the choice of the best means is no guarantee that the end pursued is the best, so the refinement of quantitative procedures cannot eliminate the irrational element involved in every political decision. Even the fact that the partners agree on the value they attach to their joint ventures does not in itself make those ventures rational. For example, by what criteria can it be said that President Kennedy's decision to put a man on the moon was rational?

> It is good that we should know what we mean [writes Raymond Aron] when we say that going to the moon is irrational. In terms of the progress of science, it is clearly irrational; in terms of economic goals, in all probability, it is irrational; in terms of national security, it is in all probability irrational; in terms of prestige, you have to measure prestige and you have to ask the President of the United States what he means by that. If he says that to go to the moon before the Russians would be a first-class victory, and that he puts the highest value on that, you may say he is crazy, but he has a case.[13]

The perfecting of budgeting techniques may give the resource allocation process more sophisticated, more objective and stricter programming instruments, but it no more constitutes planning, in the sense of social rationality, than it replaces the political process whereby conflicts of interests are settled by negotiation—or by unilateral decision.

The limitations of these techniques are all the more obvious when one tries to apply them to scientific research, The aim is still 'to reduce the area of the unpredictable to a manageable series of clear alternatives',

12. *Ibid.*, 2.
13. Raymond Aron, 'Applying First Principles', in *Decision Making in National Science Policy*, a CIBA Foundation and Science of Science Foundation symposium, Churchill, London (1968), 288.

but this unpredictability is inherent in and consubstantial with the very nature of the activities which it is sought to plan. This is not to say that scientific activities should be regarded as so specific and original that they have no counterpart in the productive system. After all, as Christopher Freeman has recalled,[14] there is a random factor in farming as well as in science, since output may vary widely from the norm in the light of seasonal variations, weather conditions, epizootic diseases and the like. The farmer is not master of the accidents of nature or the researcher of the random advancement of discovery, however much the planners may act as though an increase in the resources allocated to either of these sectors must necessarily mean greater output and a cut must mean less output.

In the case of science, however, the planning effort is subject to this special constraint: that it must set itself the *unforeseeable* as an object. Whereas, in the case of agriculture a deviation from the norm involves no surprising change in the nature of the product expected, in the case of science the little more or the little less makes the whole difference between all and nothing; no discovery or invention is, strictly speaking, inevitable. The special role of economic forecasting is to look for ratios between certain variables which have proved constant in the past and from that to infer the constancy of these ratios in the future. To take the example of national income, the economist can, in a certain manner, calculate it for future years on the assumption of a constant growth rate. But it is precisely this postulate of constancy which is lacking in scientific research; there are no statistical series of discovery, invention or scientific and technical breakthroughs which lend themselves to this kind of extrapolation.

Think of the discovery and isolation of penicillin, the classic example of the role which chance—or luck—can play in the process of discovery. Pasteur, it is sometimes said, could have identified the principle of action of *penicillium* some fifty years or so earlier, if his experimental facilities had been less meagre and if he had not been absorbed in other problems; if Cleopatra's nose had been a little shorter. . . . The conjunction of fortuitous circumstances which finally determined the discovery conforms so perfectly to Cournot's definition of chance—'the combination of events forming part of series independent of each other' —that it affords an unanswerable argument against any deterministic theory of scientific discovery: with a more modern laboratory, the

14. Christopher Freeman, 'Science and Economy at the National Level', *in Problems of Science Policy*, OECD, Paris (1968), 59–60.

culture slides that Fleming was studying might not have been contaminated by *penicillium*, which is indeed somewhat rare in the air; then, this contamination was favoured by the type of cultures Fleming had chosen, colonies of staphylococci, which are particularly sensitive to the antibacterial action of this fungus. The imponderable element that dominated the discovery of the principle of action of penicillin was continued in the isolation of the antibiotic; the product obtained by Florey was so impure that the 99 per cent of impurities which it contained would have produced toxic effects masking the therapeutic effects of the fungus; and if Florey had experimented with guineapigs instead of mice, the effect obtained would have been exactly the reverse, since penicillin is a deadly poison to guineapigs.[15] The planning of research on antibiotics could not have started before the isolation of penicillin. It was only after that, and with the aid of the second world war, that this research effort was the subject of programming designed to cut down the interval between discovery and large scale development.

If science policy does not offer 'a manageable series of clear alternatives' it is because the field remains, by definition, that of the uncertain. The objectives of free research cannot be clearly identified; *a fortiori*, the time needed to achieve them cannot be precisely fixed. Even oriented research, however meticulously it may be programmed, is never certain of the time it will need to reach its goals. Thus, the programmes on controlled thermonuclear fusion, launched in the United States, USSR and England almost simultaneously during the 1950s, were scheduled for completion in five years. Even today they are still far from reaching success.[16] The prestige of certain words should not blind us to the transference of concepts that they disguise; research programming should no more be confused with science planning than the certain with the uncertain.

No demonstration of the most deterministic conceptions of invention has yet succeeded in contradicting the words of the economist Jacob Schmookler:

All that our present knowledge permits us to say is that the probability

15. For this example and others, see René Taton, *Reason and Chance in Scientific Discovery*, Hutchinson, New York (1957).
16. Not only were the technical difficulties underestimated but, above all, the theoretical problems linked with plasma physics were ignored in the programming of projects (*cf.* H. Roderick, 'Fundamental Research and Applied Research and Development', in *Problems of Science Policy*, OECD, Paris (1968), introduction, 91).

that any given invention will be made varies between zero and one in-
clusive! [17]

The idea of optimum distribution of resources among the different research activities presupposes that it is possible not only to increase this
possibility for a given research sector, but also to measure the advantages and disadvantages of spending money in one sector rather than
another. Even at the level of a firm, where the risks involved in research can to some extent be controlled—since the calculation of probabilities defines the limits within which they are worth running or not—
this calculation of the particular advantages of different projects or
alternatives does not imply mastery of the hazards of the process of
scientific creation; the risk here is always present under its other name
of uncertainty. On the national scale it is clear that such a calculation is
even less accessible, and it is hard to see how the perfecting of quantitative planning techniques could improve matters.

The techniques for the rationalisation of decisions have all sprung
from the experience of programming modern arms systems, where the
objectives and the time set for achieving them must be defined with the
utmost strictness. The strategical stake is clear and it is given once and
for all; it is a matter of life or death.[18] If these techniques do not apply,
or apply very awkwardly, to research activities, it is not so much because
the stake is different—competition and the search for profit may determine a private firm to outdistance the others technologically for much
the same reasons as the strategic and diplomatic reasons which motivate
a nation—as because the whole nature of these activities differs from
military operations; their objectives, just like their presumed results,
are diffuse and open to challenge at any moment. One would hesitate
to advance such a truism if one did not find, in discussions on science
policy, the fascination which these techniques have for certain scientists as well as for certain administrators. Systems analysis, PPBS,
cost-effectiveness, trees of relevance and the rest, this whole arsenal of
quantitative methods is regarded as a miraculous procedure which can

17. J. Schmookler, *Invention and Economic Growth*, Harvard University
Press, Cambridge (1966), 215. On the determinist ideas of invention, see
especially S. C. Gilfillan, *The Sociology of Invention*, Follet Publishing Co.,
Chicago (1935), and above all, R. K. Merton's article 'The Role of the Genius
in Scientific Advances', in *New Scientist*, London (2 November 1961), 306.

18. 'It is misleading to say that primacy in military research and development can give us *only* lead time. This may be enough to prevent or "win" a
war, and, for a nation on the strategic defensive, it is essential to avoid defeat.'
(Charles J. Hitch and Roland N. McKean, *The Economics of Defense in the
Nuclear Age, op. cit.*, 245.)

be transposed with equal success from the realm of defence economics to that of research economics.

In reality, these techniques are no more than tools of programming and management which can serve to calculate the cost and the technical stages essential to achieve a given product or process; they are so ill-fitted to research and development activities that their underlying principles must in this case be inverted. The authors of *The Economics of Defense in the Nuclear Age* have themselves emphasised that:

> In exploratory development and research, the precise identification of objectives and scheduling are less important than trying to cover all good bets, selecting first-rate scientists and productive laboratories, promoting competition, and preserving flexibility to follow up vigorously on breakthroughs.[19]

This means, in short, that in this instance *the most effective alternative is not necessarily that which seems, at first sight, the most economical*. Some of the consequences which the authors draw from the uncertain character of research may appear to be so many challenges to the effort at rationalisation which is the aim of planning. They certainly run counter to the principles of economy which are the concern of all public administration, and which are so often invoked by the science policy-makers to 'rationalise' the allocation of resources to science. Since the result of research activities is uncertain, 'duplication of effort' —their pursuit along different lines—is all the more desirable; the more important the result expected from research, the greater its uncertainties, and the greater the duplication needed.

No doubt these principles were formulated for military research and development, in contrast with other sectors of defence economics, and the example proposed, that of the 'Manhattan Project' where six distinct and independent methods of separating fissile material were followed simultaneously, cannot be transposed into the field of civil research unless the institution concerned (firm or nation) knows the value it attaches to obtaining the result. But it can clearly be seen from this that science planning depends less on the techniques for attaining an objective, however sophisticated they may be and whatever their progress, than on *the actual setting of the objective, that is to say, the priority assigned to it in comparison with other possible objectives*. The orientation of the scientific research system will not become any more rational as a result of the application of these techniques to the execution of programmes, however great the improvements introduced into their

19. *Ibid.*, 249.

administration. Few projects, after all, have been more meticulously programmed than the 'Apollo Project', more strictly subjected to systems analysis, and conceived precisely as one of the most complex undertakings which a research administration has ever had to handle. The rationalisation methods which made it possible to solve the economic and technical problems that had to be faced, once the project was decided upon, do not thereby afford any guarantee of the rationality of the project itself.

Technological forecasting

And yet, it is from faith in these methods that the idea has germinated of a 'policy technology for science', capable of scientifically overcoming the uncertainties of research and, in consequence, of planning science with quantitative formulæ at least as coherent and effective as those on which economic planning is nourished. Once again, it is not that recourse to these methods as management tools is not imperative and does not introduce greater enlightenment, and therefore greater effectiveness, into the process of making and implementing decisions. It is commonplace to say that the perceptible quickening of change and the growing complexity of their management systems compel modern societies to project themselves into the future.

> On a familiar road [said Gaston Berger, pleading long ago for the prospective approach] a driver travelling by night at a walking pace needs no more than a feeble lantern, but a powerful car speeding through strange country needs powerful headlights.[20]

The very success of certain long-term forecasts reflects the curiosity of the general public for an apparently new formula quite as much as their apprehensions for the future, a future which is all the more disturbing since the speed and accumulation of technical progress seem to make it more imminent and more inevitable. But if forecasting is recognised as a necessary discipline to the point of becoming an industry in its own right,[21] it does not follow that its procedures have acquired the character of scientific demonstration, or, especially, that it eliminates or even diminishes the ambiguous texture of decisions. The art of

20. Gaston Berger, 'La Prospective' (1957), in *Phénomenologie du Temps et Prospective*, Paris, P.U.F. (1967), 221.

21. The annual investments of American enterprises in R & D institutes and firms are valued at more than than $65 million for technical forecasts alone. See Erich Jantsch, *Technological Forecasting in Perspective*, OECD, Paris (1967), 251–3 and 272.

conjecture, however strict the scientific apparatus on which it rests, remains an art.

This does not mean, either, that science policy can do without a certain amount of forethought. Research programmes do not conform to the annual pattern of national budgets and their results can never be accounted for in the short term. Since 'government is foresight', research administration must inevitably project itself into the future. Even more, when we think of the consequences, direct and indirect, unexpected or undesirable, of technical progress, what field of political action has greater need of foresight than this? Since the applications of science and technology provide the most powerful instruments of change, it seems all the more necessary to influence the course of events by orienting technical progress in the light of its implications rather than of its genesis. But while we can (up to a certain point) predetermine the nature, number, and even the timetable of new technologies, we are in no position to foresee their effects on social evolution.

To take the very simple example cited by Bertrand de Jouvenel, the decline in domestic service in the industrialised countries has been the consequence not so much of technological progress as of a deliberate policy of full employment, and if in the United States, in a period of underemployment, the shortage of domestic staff is not relieved when the numbers of unemployed go up, this is because domestic service has no 'psychological attraction', and all the more because there is 'the alternative resource of unemployment benefits, resulting from political measures'.[22] The technical revolution in household appliances did not do away with domestic service any more than the invention of the horsecollar in the west did away with slavery. There had to be the roundabout route of political initiatives which had nothing to do with technological progress (the multiplication of household appliances itself being a consequence of growth linked with full employment, rather than of their technical improvements).

> If, in 1913 [said Bertrand de Jouvenel], you had 'fed' a social forecaster with the whole of the technological evolution of the next half-century, he would never have inferred from it the disappearance of domestic staff.[23]

Utopias, too, depicted the future countenance of a society, a future more or less plausible, depending on the links they preserved with the conditions of the present. Unfettered by time, they pictured a future

22. Bertrand de Jouvenel, *The Art of Conjecture*, Weidenfeld, London (1967).
23. *Ibid.*, 358.

at some imaginary date, in the form sometimes of a return to the 'Golden Age' of primeval perfection, and sometimes of a mythical image of a metamorphosed present.[24] The prospective approach, in contrast, does not evade time by restoring or founding the ideal city; it proposes dates, milestones, promises, which fall into a mathematical form, constitute a timetable and pinpoint the due dates of 'events that shape the future'. But if the prospective approach challenges chance on the same ground on which the utopias rejected time, the imagination of the future has retained the same function of exorcism, and the spatial language of the probable affords no more rigorous long-range vision than the atemporal language of preferences. Social forecasting is based on the projection of trends linked with variables whose number, though greater and more sophisticated than those which nourished the utopias, nevertheless remains limited. Furthermore, as economists have long known, the limitations of all forecasting are set less by defects of method than by the absence of adequate data about the present, uniform, full and, above all, available at the moment they are wanted:

> Getting the data in time [recalls Donald A. Schon] may be as critical as getting it at all, as when we attempted in 1965 to make forecasts to 1975 on the basis of the 1958 data.[25]

It is no different when the prospective approach is applied to scientific research. In the first place, the process, the institutions and the performers of scientific research constitute such a complex system that it is not feasible to assemble and master all the information that seems necessary. Secondly, this system itself cannot be treated as if it had a life of its own, independent of all other systems. As Gilbert Simondon has emphasized, the technical object has a defined mode of existence because it has a genesis, but this genesis is not only a genesis of objects, it is the object's history in its relation to man and to the world.[26]. Technological forecasting defies chance as though it were dealing with a reality independent of man, peopled with pure objects whose probable existence is defined by the instrumental model of the finished object, disposable and manageable, the tool detached from its history rather than integrated in the generic terms of its relation to man. And yet that relation is there from the start of play; the pinpointing of the

24. See *Les Utopies de la Renaissance*, P.U.F., Paris (1963).
25. Donald A. Schon, 'Forecasting and Technological Forecasting', in 'Toward the Year 2000: Work in Progress', *Daedalus*, Boston (summer 1967), 765.
26. Gilbert Simondon, *Du Mode d'existence des objets techniques*, Aubier-Montaigne, Paris (1969), especially 154–8.

utopia does not make the genesis of the technical object a game without human partners. Technological forecasts are never more than conjectures as to trends, that is to say, the result of a series of questions put to experts, whose views, however they may be mathematically processed, do not thereby cease to be a matter of opinion.[27]

Erich Jantsch's book is the one which has done most to propagate among national administrations the idea of technological forecasting as an instrument. It is a fascinating book in many respects, not least in the panorama it affords of the many methods and institutions devoted to 'what is defined as the probabilistic assessment, on a relatively high confidence level, of future technology transfers'.[28] The concept of 'technology transfers' endows forecasting with a new dimension in that it proposes a 'space' within which the ambition is to forecast discovery, invention and innovation, not only in themselves but also in so far as they have repercussions on the social environment. The first type or level of forecasting is content to determine the time needed to perfect an invention, the efforts needed to obtain it and its functional possibilities. This is what Jantsch calls 'exploratory technological forecasting' as distinguished from the higher level of 'normative technological forecasting' in which the consequences of the invention themselves must be the subject of forecasting. The first level, in short, is concerned with the history of the invention itself, the second with the way in which the invention should affect history in general.

In a certain sense, all the literature stimulated by the prospective approach seems to smack of *The Morning of the Magicians*, Erich Jantsch's book no less than the others.[29] But the sense of responsibility

27. The 'Delphi method' is the best known example of procedures which do not conceal the fact that they are based primarily on intuition. Carefully programmed questions on the technological trends in a given field are put to a group of experts, and their replies collated with information obtained from outside; these 'brainstorming' sessions are then computerised. The result is a table of forecast technological breakthroughs, following a timetable composed of intervals of probabilities arrived at by reasonable agreement among the experts. As for technological forecasting methods which try to reduce the role of intuition by the use of different mathematical tools (matrices, simulation, scenarios, input–output tables, etc.), it is obvious that they cannot do without preliminary enquiry from experts. On the 'Delphi method', see T. J. Gordon and Olaf Helmer, *Report on a Long-Range Forecasting Study*, Report, Rand Corporation, Santa Monica (September 1964), 2982: on the other methods, see Erich Jantsch, *Technological Forecasting in Perspective*, OECD, Paris (1967), second part, chapters 3 and 4.

28. Erich Jantsch, *Technological Forecasting in Perspective*, *op. cit.*, 15.

29. For example, when it places on the same level methods whose degree of assurance is very different, or treats facts as established which are not. Thus, one

of the institutions whose activities he lists and describes, and indeed of
the international organisation which sponsored his book, is enough to
ensure that he is not to be taken lightly. The optimism he displays
towards the possibilities of technological forecasting is a revelation of
the spirit in which science policy is steeped; many are the administrators who share this optimism and who are ready to believe that, provided techniques are perfected and applied systematically to
government decisions affecting science, rationality will soon make a
mockery of history by mastering all chance events.

From this point of view, technological forecasting seems nearer to
the Roman technique of interpreting omens than to the mythical image
of the ideal city in the utopias of the Renaissance. For these utopias, in
so far as they reckoned nothing of time, had no concern with immediate
action; the augurs aimed at mastering the signs which determined
action.[30] Just as it was the function of the augurs to 'valorise', in the
primitive meaning of the term, 'the omen in itself, become, by virtue
of the rite, decisive',[31] so the technological forecasters valorise the lines
of force of science and technology, destined by virtue of mathematical
calculation to become in their turn decisive. The rational and methodical approach transfers the interpretation of signs from the unfathomable
terrain of nature to the area of culture, capable of being reduced to form.
The message of the future is no longer to be deciphered from natural
manifestations in which the gods may perhaps intervene between man
and his *fatum*, but from the products of culture in which the future
of technology is matched with destiny. But the issue has not changed; it
is still immediate action, the choice to be made at the moment, the
decisions to be taken *hic et nunc*, which are to be inspired.

And yet it is easy to question the assurance of the technological forecasters, since even the first level of their conjectures, that of 'exploratory
forecasting', is still far from having proved itself—except again as a
tool for programming the transition from the stage of completed discovery or invention to the stage of innovation. The efficient working of

---

reads that 'only two countries have so far established a framework in which
technological forecasting can be used systematically to aid national planning:
France and the United States'. It is not the 'only' which shocks, but the assertion that such a framework has ever existed in France or the United States.
E. Jantsch, *Technological Forecasting in Perspective*, op. cit., 279.

30. Jean Bayet, *Histoire politique et psychologique de la religion romaine*,
Payot, Paris (1957), 51–60, and Raymond Bloch, *Les prodiges dans l'Antiquité
Classique*, P.U.F., Paris (1963), part 3.

31. Jean Bayet, *op. cit.*, 102.

discoveries which is ensured by modern management techniques induces the belief that, by the same mechanistic process, discovery itself can be reduced to the execution of a programme, since forecasting methods allow the potentialities of discovery to be transformed into possibilities, or in other words allow the image of the discovery conceived in time to be treated mathematically as a state already achieved in space. But control of the process by which the product of scientific research is obtained is still not the realised product. However clearly it may be visualised in its technical affiliations and its economic restraints, the uncertainties of its genesis are not thereby swept away.

This could of course occur if it is maintained that the discovery to come, like all the future work of which Bergson spoke, is locked away 'in some box or other, full of possibles' whose key the experts, 'by virtue of their already long-standing relationship' with science and technology, will wrest from the forecasts.[32] The illusion is just the same in speculations on technical change as in metaphysical speculation, even if the event seems more easily calculable as a future state of a closed system of material points; this is to blind oneself to the fact that it is abstracted from a totality which does not include merely space and matter. 'If I knew what would be the great dramatic work of tomorrow, I would create it'. One of the best specialists in technical innovation has an answer no different from that of Bergson:

> There is a special problem about any theory that presumes to permit prediction of invention [writes Donald A. Schon]. In one sense, a prediction of invention *is* invention, and the prediction fulfils itself. To claim that a theory permits prediction of invention is to claim that a theory permits invention.[33]

But it is when we move on to the higher level of 'normative technological forecasting' that the technical forecasters reveal with less ambiguity the role of augurs which they play, or want to play, in modern societies. 'Normative technological forecasting' postulates in effect that it is possible, by analysing the parameters of the discovery to come, to foretell its applications, the needs they will satisfy and the effects they will have on the social environment. The 'space of technology transfer' is presumed to be explorable in the totality of its future, like a feedback system where the knowledge of the result, at once the cause and the

32. Henri Bergson, 'Le Possible et le Réel' (1930), in *La Pensée et le mouvant*, P.U.F., Paris (1950), 110.

33. Donald A. Schon, 'Forecasting and Technological Forecasting', *Daedalus*, loc. cit., 767. See his *Technology and Change*, Delta Books, New York (1967).

effect of the decisions of which it is the product, in all probability allows it to be achieved.

The art of forecasting here tends to be substituted for the art of decision; it transfers the promises induced from technological trends into the realm of social objectives. It is 'normative' precisely in that it founds its conjecture on the technical objective, on the values by reference to which the achievement of that objective should be decided upon. One might almost believe that Norbert Wiener's hopes had been fulfilled for technical knowledge grown conscious of its own ends; does not 'know-how' extend itself into 'know-what'?[34] But the profile of the future, modelled from those technical family trees which are all that intuition, even with the aid of mathematical models, can forecast with maximum plausibility, is thoroughly steeped in today's values; the options inspired by tomorrow's technological imagination are inseparable from today's ideological realities.

No doubt, like the Roman augurs, the technological forecasters stress that they confine themselves to designating alternatives and that their prophecies can always be rejected—*omen exsecrari*. The Romans became so skilful in this art of reassurance about the scope which the vision of the future leaves for freedom of action that, by mastering the techniques to which they subjected fate, they gave the impression of mastering fate itself. Thus:

> These technicians themselves, faithful to the tendency of the Latin spirit, became more and more the masters of the signs to which they were deemed to be subject . . .

—to the extent of controlling the appetite of the sacred chickens in their cages.[35] Similarly, technological forecasts must have propaganda value if they are to be fulfilled; the possibilities which they leave to chance are all the less the greater the weight of talk about the future in the decisions of the present. The lines of force, the trends and directions of scientific research, are designated as probable on the basis of present values, which make them desirable and at the same time invest them with a higher probability coefficient.

> The important point to remember [says one of the champions of 'normative technological forecasting'] is that a planning system thus expanded

34. 'There is one quality more important than *"know-how"*, and we cannot accuse the United States of any undue amount of it. This is *"know-what"*, by which we determine not only how to accomplish our purposes but what our purposes are to be.' Norbert Wiener, *The Human Use of Human Beings*, Avon Books, New York (1967), 250–1.

35. Jean Bayet, *op. cit.*, 55.

permits the introduction of objectives and policy-goals as part of the anticipation, and these become operational elements in defining the changes that are needed in the present—in suggesting the sets of policies that need to be applied, the interrelated and interactive policies that need to be pursued—if the anticipated preferred future is to be translated into current reality.[36]

The heuristic function of the modern augurs is no less political than the function of divination among the Romans. Just as that divination became a tool in the hands of authority or candidates for authority, assigning success or failure to their enterprises in advance, so 'normative technological forecasting' nurses the ambition to be at the centre of state decisions, determining their orientations and their ends. 'How is an *a priori* history possible?' asked Kant ironically. Answer: 'If the seer himself *creates* and arranges the events which he foretells'.[37]

Forecasting and free research

In holding itself out as the instrument of a social technology,[38] technological forecasting is not satisfied with interpreting the signs which indicate the possible directions of scientific research. It commits itself to a conception of research and of society in which the search for new knowledge is concerned only with the applications which that knowledge makes possible. The 'normative' horizon of technological forecasting is utility, whereby the discovery and the invention are translated into innovation and affect not the relation of the researcher to the object of his research, but the relation of the research product to the social whole. Thus, forecasting techniques are not only techniques, but also ideologies whose postulates set the frame within which the research system is assigned not only its subject but its ends.

Free research, the results of which remain uncertain for the researchers who undertake it and the institutions which support it, appears to be an extreme case—a deviation from the postulates of utility and economic return—only so far as it is admitted that the research system includes an obscure and irreducible zone, doomed to chance and surprise. It is therefore enough to postulate the total integration of the

36. Hasan Ozbekhan, *The Idea of a 'Look-Out' Institution*, System Development Corporation, Santa Monica, California (March 1965), quoted in Jantsch, *Technological Forecasting in Perspective, op. cit.*, 244.

37. I. Kant, *The Conflict of the Faculties*, in the pamphlets on *The Philosophy of History*.

38. Olaf Helmer, *Social Technology*, Basic Books, New York (1966), and Hasan Ozbekhan, *Technology and Man's Future*, report SP-2494, System Development Corporation, Santa Monica, California (27 May 1966).

research system in the social whole, free research included, even with a lesser degree of assurance than for the other forms of research, and the whole difficulty is overcome. Predetermining what it must aim at, and above all what it must be used for, the forecasting technician will be in a position to direct free research in line with the social aims of which he is the augur.

> Normative technological forecasting [writes Erich Jantsch], starting from social requirements, is capable of applying spur and guidance to fundamental research in areas of social relevance, in the same way as they are applied by industry in the economic area.[39]

Jantsch cannot be charged with failing to put the question fairly and squarely. 'Normative technological forecasting' challenges the whole idea of a 'pure' science with characteristics of such a kind that its evolution cannot in any way be foretold. This is the theory of what he calls the 'encapsulation' of science—its withdrawal into an ivory tower, immune from the pressures of the profane world—and of which he finds, not without reason, one of the best examples in the book by Thomas S. Kuhn, *The Structure of Scientific Revolutions*. According to Kuhn, scientific progress is made up of two sorts of movement, that of 'normal science' which develops within the limits of established 'paradigms', and that of science in a period of crisis, when the revolution set off by the 'anomalies' of the concepts in use takes the form of strife between the old and new 'paradigms', until the victory of the new concepts, recognised and adopted, gives rise to a new 'normal' science'.[40]

The paradigms provide a criterion which is enough in itself for the selection of the problems to be solved. If the scientific enterprise proves useful, it is because of the solutions provided in the context of 'normal science'. But it is impossible to influence this process from outside, and *a fortiori* to foresee the 'anomalies' which become the source of new paradigms.

> We are deeply accustomed [says Kuhn] to seeing science as the one enterprise that draws constantly nearer to some goal set by nature in advance.[41]

39. Erich Jantsch, *Technological Forecasting in Perspective*, *op. cit.*, 60; see also the same author's 'Technological Forecasting—A Tool for a Dynamic Science Policy', in *Problems of Science Policy*, OECD, Paris (1964), 113–23.

40. Thomas S. Kuhn, *The Structure of Scientific Revolutions*, Chicago University Press (1962). In many respects Kuhn's anomalies recall the 'epistemological obstacles' of Bachelard in *La formation de l'esprit scientifique* (Vrin, Paris (1938), 91). But just as the idea of 'paradigm' is vague, so that of the 'epistemological obstacle' is precise and rich to the point of being the principle which explains the 'anomalies' themselves.

41. Kuhn, *op. cit.*, 170.

It does not help, however, to imagine that there is a single full, objective, true account of nature, and we should

> ... account for both science's existence and its success in terms of evolution from the community's state of knowledge at any given time. ... If we can learn to substitute evolution from what-we-do-know, for evolution toward what-we-wish-to-know, a number of vexing problems may vanish in the process.[42]

This assuredly 'purist' conception, which rejects all influence over the course of science other than that of its own problems, contrasts at the opposite extreme with the conception of the 'integration' of science in the social system. The empirical course of history takes precedence here over the theoretical consideration of knowledge, as though knowledge had no significance except in so far as it is conditioned by history. *If these conceptions are diametrically opposed, it is not so much because they both alike find illustrations founded on the facts, but rather because they each refer back to irreconcilable ideologies.* We have a dialogue of the deaf, because each camp refers to an objective which could be defined independently of the values it attaches to the objective.

> Both positions [said Georges Canguilhem] come down to treating the subject of the history of sciences as the subject of a science.[43]

Hindsight or foresight, the misapprehension is the same; if one finds here and there, formulated in terms as absolute as those of the 'idealistic' interpretation, the conception of the total integration of science in society, it is because the subject of reflection is in both cases presumed to be determinable *as the subject of a science.* The idea of a 'pure' science, sheltered in its ivory tower to the point of being heedless of the noise of the modern world, is challenged by the evolution of the relations between knowledge and power, contradicted by the intervention of power and the dependence of knowledge, transformed by the absence of strict frontiers between the different stages of the research system. But, just as the social demands and responses of which science is the

42. *Ibid.*
43. See Georges Canguilhem, *Etudes d'histoire et de philosophie des sciences,* Vrin, Paris (1968), introduction, 15, referring to the dialogue between Alexandre Koyré and Henry Guerlac at the conference in Oxford (July 1961), where Guerlac accused Koyré of being an 'idealist', that is to say, of regarding scientific activity purely as a theoretical activity and giving to the facts of the history of sciences a reality independent of the social context. (Henry Guerlac, *Some Historical Assumptions of the History of Science,* reprinted in *Scientific Change* (ed. A. C. Crombie), London (1963), 797–812, and the reply of Alexandre Koyré in *Etudes d'histoire de la pensée scientifique,* P.U.F., Paris (1966), 352–61.)

subject do not mechanically explain the roads it follows, so the prediction of its possible results does not mechanically determine the processes needed to achieve them. The ambition displayed by the technological forecasters to direct science—and society—on the basis of a scientific forecasting of its possible directions does not do away with the coefficient of uncertainty attached to discovery and invention. And indeed the utilitarian postulate, on the basis of which the future of science, brought into relation with social needs, is deemed to be predeterminable, does not lead to the application of techniques which are, or even which can be, mastered.[44] It is impermissible to pass off conjectures on the future of the scientific enterprise as being fully rational and to treat them as the scientific foundation of possible political decisions; this scientific foundation is non-existent, and conjectures do not cease to be conjectures by relying on mathematical instruments rather than on intuition.

It is true that there are some 'strategic' decisions which take the form of a more rigorous determination of the fields in which, in view of the results expected, efforts should be concentrated. The country which wishes to equip itself with nuclear power must necessarily have researchers capable of solving the problems of fission and the perfecting of reactors; or again, since molecular biology is clearly a science 'of the future', similarly spectacular results may be expected from DNA in the next decade to those which followed in physics, around 1930, from the discovery of the neutron. But this determination is in no way an objective and rigorous discourse on the technical possibilities of the future and their repercussions on the social whole. From the outset, even if masked by the mathematical apparatus, it is surrounded by the values in which its authors are steeped.

> One may even venture to predict [writes Jantsch] that technological forecasting will be largely instrumental in determining fundamental research in the near future.[45]

44. Jantsch, moreover, contrasts Kuhn's views with those of R. G. H. Siu, who proposes, in *The Tao of Science*, the addition to western knowledge founded on reason of certain elements of eastern wisdom, particularly the idea of a 'non-knowledge' capable of piercing the mysteries of nature (*op. cit.*, 59–61). Scientific creation would be 'the fluorescence of non-knowledge', neither rational nor intuitive, but a communion. There is no better illustration than this reference of the parascientific and divinatory terrain of technological forecasting; one could ask what practical advantage the technological forecasters would be able to derive from it. See R. G. H. Siu, *The Tao of Science*, MIT Press (1964), especially chapter 9.

45. *Op. cit.*, 57.

One may no doubt venture, but if fundamental research is likely to be increasingly directed, it is not because the forecasts in whose name research is directed will have any greater force of probability tomorrow than they have today, it is because what they will determine as probable will be increasingly merged with what they deem, implicitly or explicitly, to be desirable.[46]

Just as in the history of sciences there is no forerunner in the strict sense of the word—because a forerunner is a man of whom it can be said that it is not until afterwards that he is known to have gone before —so technical forecasting does not predict, in the strict mathematical and deterministic sense of the word, the scientific and technical breakthroughs of tomorrow, or *a fortiori* their effects on the social environment. It can serve as a guide only with 'all other things *becoming* equal', that is to say, at the risk of the very factor which it professes to eliminate—the gamble on the future.

> The subject matter of the historian of the sciences [says Georges Canguilhem] can be delimited only by a decision which assigns it its interest and importance.[47]

In the same way, the subject matter of technological forecasting is circumscribed by the decisions of the technological forecasters who apply to it their own frame of reference. If technological forecasting is 'normative', this is not because it finds ready-made values stored away in science's 'box of possibles', implied in advance in some formalisable relationship between applications and social needs, but because it colours the scientific future, to which it professes to hold the key, with its own values. The normative space of the forecast is not the objective, logical, rational space of scientific thought. In spite of the mathematical apparatus with which it surrounds itself, it is never more than one of the elements of the equivocal space in which decisions are reached and applied. If the prophets advise the prince, they can be sure of their vision of the future only if they themselves make it happen.

46. Relying on the conclusions of 'Project Hindsight', Jantsch writes, '*The absence of normative thinking* had been found to render fundamental research quite useless for the purpose of American defence developments'. (Author's italics; *Technological Forecasting, op. cit.*, 54.) In the first place, it is not true, and secondly and above all, it is clear that the definition of the useful becomes the pretext here for the retrospective definition of the probable. On 'Project Hindsight', see C. W. Sherwin *et al.*, *First Interim Report*, Office of the Director of Defense Research and Engineering, Clearing House for Federal Scientific and Technical Information, Washington (1966).

47. Canguilhem, *op. cit.*, 18.

CHAPTER 6

Science Policy and its Myths: The Allocation of Resources

The master myth cherished by the bureaucrats of science policy is that forecasting can be applied to science as an instrument of planning to set the future course of scientific research. The act of decision is treated as though it were an integral part of the rationalised world of technology and as though it were itself only one technique among others. Since science consists of forecasting, the planning of science must meet with the same 'operational' success as all techniques based on rational knowledge.

Administrators and politicians are not alone in cherishing this dream; in the corridors of power the whole idea of a 'social technology'— mainly encouraged by the technological forecasting specialists—fits into the dreams of intervention and of manipulating the social sciences when their subject matter is treated as being on an equal footing with the subject matter of the natural sciences. The idea is that the language of culture can be deciphered in the same mathematical terms as the language of nature, and expresses in the same terms the transition from formalised thought to empirical action. In short, it would be enough to tailor from the experience of social phenomena the same 'well-fitting garb of ideas', as Husserl calls it, which has proved itself in the experience of physical phenomena, the garb of mathematical symbols, theories and truths, to ensure that the functioning of the social system is reduced in its turn to the operation of a machine.

There is no reason to think that the gap between the rational and the real is any wider in science policy than in other fields of policy, even if the mind boggles more at finding that the rational refuses to yield to reason in this way. What should really cause more surprise is the shock of finding that science policy, on the pretext that it relates to the affairs of science, is expected to prove itself more scientific and more rational, more capable of minimising the element of historical contingency, of the unforeseeable and uncontrolled which is the badge of all human undertakings, than policy in other fields into which rationality does not enter either as a requirement or an aspiration. The rationality which is the pride of science impregnates all discussions on science policy, as though science policy decision-makers could expect to benefit from greater rationality than any other political decision-makers. But the theatre in which political decisions are acted out, with its actors, its stage managers and its comings and goings, does not cease to be historical by the mere fact that its theme is science. The universe of politics does not lend itself to formalisation, in spite of the efforts of certain political scientists and the techniques they apply to mould the contingencies of history into the terms of scientific analysis. As Raymond Aron recalls in connection with 'games theory', we cannot expect political science to be an 'operational' science in the sense in which physics, or even certain parts of economics, are operational.[1]

Decisions are nevertheless taken and choices are made, the criteria for which are based on some form or other of rationality and which can therefore, at any rate in theory, be made explicit, if not improved. For the scientists who represent the strategic reserve of science policy, it is not immaterial to know the level of rationality at which decisions are taken affecting them, which, whether they like it or not, deploy them on the battlefield. But for those scientists who are promoted from the ranks of combat units to serve on the headquarters staff and find themselves taking part in the decisions, this question becomes a real acid test; can they, by mastering the relevant facts, infuse greater rationality into decisions affecting science by virtue of their greater ability than that of other men, especially politicians, to bend the facts to the rules of the scientific method?

This is the underlying significance of the debate opened in the journal *Minerva* by Alvin Weinberg, which is worth following point by point in order to learn its limits. It is a purely academic debate; there can be no question of these scientists stepping into the shoes of the political decision-makers, but only of their throwing light on the actions of these decision-makers on the basis of rules of the game so methodically defined and so rationally founded that the 'research community', in the very name of science itself, finds no difficulty in bowing to them:

> Thus, as a practical matter, we cannot really evade the problem of scientific choice. If those actively engaged in science do not make choices, they will

1. R. Aron, *Peace and War—A Theory of International Relations*, New York (1966).

be made anyhow by the Congressional Appropriations Committees and by the Bureau of the Budget, or corresponding bodies in other governments.[2]

Since in any event a choice has to be made, the scientists are caught in the trap of governing men instead of handling objects. Thus, shifting the problem from the theatre of politics on to the ground of scientific deliberations, they will try to formalise the terms of the relation which inevitably binds science to power or, in other words, they will try to reason about science as a means without questioning the ends it serves.

The criteria of choice

At the level of the researcher or of the institution in which he works, decision, in theory, need be embarrassed by no criteria except that of the quality of the men and of the projects; their scientific value, recognised and confirmed by 'judgment of their peers', is an unchallengeable warrant for the support they demand. In this sense academic research certainly seems the nearest version to the theoretical model of the market economy. The republic of science, says Michael Polanyi, is made up of the self-coordination of individual scientific efforts, as though they were guided by the 'invisible hand' which Adam Smith invoked to describe the self-adjustment of relations between producers and consumers.[3]

This would be an attractive analogy if it were only the content of research that was at issue; no outside intervention, of course, will, in the eyes of any researcher worthy of the name, confer scientific value on a piece of mediocre research work or an absurd project. It nevertheless sometimes happens that a young researcher who presents a 'sound' research project is not supported by his elders. The authority of scientific judgment may be harmful in so far as it becomes, even on the scientific plane, an authoritarian judgment. The inertia of university institutions and the resistance of 'vested interests' have never favoured new research at the interface of several disciplines; conversely, neither 'bogus science' nor 'false sciences' can keep going for very long among men and institutions whose scientific standing is beyond challenge.[4]

2. A. M. Weinberg, 'Criteria for Scientific Choice', in *Criteria for Scientific Development: Public Policy and National Goals—a selection of articles from 'Minerva'* (ed. E. Shils), MIT Press, Cambridge, Mass. (1968), 23.

3. M. Polanyi, 'The Republic of Science', *ibid.*, 2–3.

4. The multiplication of committees and referees will never provide sufficient guarantee against collective error and the repetition, for example, of an adventure such as that of the 'N rays'. See, on this affair, Jean Rostand, *Science fausse et fausses sciences*, Gallimard, Paris (1958), chapter 1, 14–40.

This means that, even at this level, the republic of science cannot function to the general satisfaction any more than the market economy can; nowhere has the 'invisible hand' ever been enough to ensure the balance of the interests at stake. In a university which finds itself allocated a block grant for its research activities, it is inevitable that the division of the spoils should arouse parish-pump disputes in committees and subcommittees; why should this department, or that sector, or the other researcher, grab a larger slice of the cake? It is even worse at government level, where non-scientific considerations inevitably encroach upon the specific values of science. Michael Polanyi's liberal model asserts, like Renan, the blind duty of society towards science, without any corresponding duty on the side of science except the truth. The slightest external intervention is enough to distort the mechanism of the free market, and the self-regulation of the network of initiatives and ideas, of which scientists are the sole judges, is the only guarantee of the progress of science: 'You can kill or mutilate the advance of science, you cannot shape it.'[5] And it is an act of faith in the interests and objectives of the body politic that he looks to it for support 'without obligations or sanctions':

> Those who think that the public is interested in science only as a source of wealth and power are gravely misjudging the situation. There is no reason to suppose that an electorate would be less inclined to support science for the purpose of exploring the nature of things than were the private benefactors who previously supported the universities.[6]

Everything in fact points the other way: the accelerated expansion of science budgets in the last twenty-five years, far from indicating that the government is acting as a disinterested patron, is an assertion of the collective concern to integrate the research system in the economic system as one force of production among others. And it is by no means certain that this support is likely to go on growing at the same annual rate, unless we imagine that research and development activities will end by swallowing up the whole of the public resources. In fact this growth rate has been tending since 1967 to slow down or level off, even in countries where it used to be three or four times as fast as growth of the gross national product. Lavish government support at that time was rather like publicly organised gold rushes to California; the gold mines are not exhausted but the period of the mad rush is on the point of coming to an end. If the government has learned a lesson, it is that not all lodes have the same attraction for it. Some are more

5. M. Polanyi, 'The Republic of Science', *op. cit.*, 9.   6. *Ibid.*, 14.

promising, others more profitable or less costly in the light of criteria
which can no longer be defined solely by the prospectors; the greater
the interest the public authorities take in science, the more selective that
interest becomes.

This is clear from the example of high energy physics, which immediately after the second world war benefited from the prestige won by
nuclear research. The expectations were different and, in truth, not very
closely related to those of nuclear physics; but political circles, from
executives to parliaments, were all willing to subsidise the most expensive accelerators, as though the practical success achieved in one sector
of physics were bound to be repeated in others by dint of budget
appropriations, giant machines and large research teams. High energy
physics, a field in which experiments involve envisaging constantly
larger installations, today heads the list of 'agonising reappraisals' everywhere. From Brookhaven to Stanford, Geneva or Dubna, so much has
been learned in the last ten years that this field seems saturated with new
knowledge and incapable of mastering the capital of information it has
accumulated. But if science is in crisis in this field because it cannot
dominate its knowledge, it is no longer important, or no longer so
important, to the government to guarantee it the means of overcoming
the crisis.

Weinberg starts by recalling that the system of scientific panels
suffers from being dependent on specialised experts who inevitably share
the same passions and the same enthusiasms: the cause of oceanography is sacred for the oceanographers and that of high energy physics
for the specialists in sub-nuclear particles, but the two causes cannot
be made commensurable in the eyes of either group. The specialised
panel is competent to judge in its own field, but precisely because it
consists of experts whose views are those of a coterie, it cannot place
proposals within its competence in their proper perspective, or, in
other words, it cannot decide their relative value for science as a whole:

> We can answer the question 'how' within a given frame of reference; it
> is impossible to answer 'why' within the same frame of reference.[7]

He therefore proposes to improve the panel system by introducing for
each specialised field representatives of neighbouring fields, representatives of low energy physics for committees competent in high energy
physics, of nuclear physics for low energy physics, and so forth. This
suggestion is a palliative only, the real question still being that of how

7. *Ibid.*, 23.

to formulate a scale of values which might help establish priorities among scientific fields whose only common characteristic is that they all derive support from the government.[8]

Two categories of criteria must be distinguished, says Weinberg: the 'internal' criteria specific to the field of science in question, which measure *how* the research is conducted—well or badly—and the 'external' criteria, generated outside the scientific field, which determine *why* a given field should be supported. Of these two categories, the second is the more important, since those in the first category can in any event always be identified by the advice of the experts who are familiar both with the field in question and with the researchers working in it; is the field ripe and ready for exploitation? Are the scientists in the field really competent? But as Weinberg emphasises, we cannot base our judgments entirely on internal criteria, since it is vain for scientists to regard the pursuit of science as such as society's highest good; that is an opinion which they are alone in believing to be generally accepted. In the United States, for example, many bright young men today prefer to go in for biological research rather than medical practice, and government support is generally available for postgraduate study leading to the PhD degree but not to the medical degree.

> It is by no means self-evident that society gains from more biological research and less medical practice. . . . Science must seek its support from society on grounds other than that the science is carried out competently and that it is ready for exploitation.[9]

Weinberg identifies three external criteria; technical merit, scientific merit and social merit. The first presents no great difficulty. If a technical objective is deemed worthy of interest, it is important to support the scientific research which makes it attainable; if we want to learn to make breeder reactors, we must first measure the production of isotopes in neutrons as a function of the energy of neutron bombardment. No doubt it is not easy to determine in what respect any particular element of pure research is technically appropriate: the utility of the laser became apparent after and not before the discovery of the principle of optical amplification. But that, says Weinberg, is an exception rather than the rule; most 'programmatic basic research' activities can be associated with a technical objective, if not in detail, at any rate in broad terms.

Scientific merit, on the other hand, is already rather less clear, since

8. *Ibid.*, 21.   9. *Ibid.*, 25–6.

it cannot be judged for a specific field without relating it to other fields. Here Weinberg cites the words of the late John von Neumann:

> As a mathematical discipline travels far from its empirical source, or still more if it is a second and third generation only indirectly inspired by ideas coming from reality, it is beset with very grave dangers. It becomes more and more pure aetheticising, more and more purely *l'art pour l'art*. This need not be bad if the field is surrounded by correlated subjects which still have closer empirical connections, or if the discipline is under the influence of men with an exceptionally well-developed taste. . . . At a great distance from its empirical source, or after much 'abstract' inbreeding, a mathematical subject is in danger of degeneration. At the inception the style is usually classical: when it shows signs of becoming baroque, then the danger signal is up.[10]

Weinberg believes that von Neumann's observation can be extended to the empirical sciences; empirical basic sciences which move too far from the neighbouring sciences in which they are embedded tend to become 'baroque'.

> Relevance to neighbouring fields of science is therefore a valid measure of the scientific merit of a field of basic science. . . . A field in which lack of knowledge is a bottleneck to the understanding of other fields deserves more support than a field which is isolated from other fields.[11]

Thus, the original motivation of much high energy physics is to be sought in its elucidation of low energy physics, and of measuring the neutron capture cross-sections of the elements in the elucidation of the cosmic origin of the elements.

> The discoveries which are acknowledged to be the most important scientifically have the quality of bearing strongly on the scientific disciplines around them. For example, the discovery of X-rays was important partly because it extended the electromagnetic spectrum, but much more because it enabled us to see so much that we had been unable to see.[12]

All other things being equal, says Weinberg, that field has the greatest scientific merit which contributes most heavily to and throws the brightest light on its neighbouring scientific disciplines.

Finally, social merit, or 'relevance to human welfare and the values of man': Weinberg recognises from the outset that this is a vague criterion and difficult to clarify. Who is to define the value of man or even the values of a given society? And how is it decided whether a given scientific or technical enterprise indeed furthers our pursuit of social values, even when those values have been identified? There is little trouble

10. *Ibid.*, 27. The quotation from John von Neumann is taken from *The Works of the Mind* (ed. R. B. Heywood), University of Chicago Press (1947), 196.   11. *Ibid.*, 27.   12. *Ibid.*

about some values such as adequate defence, or more food, or less sickness; in other cases the difficulties of clarification seem insurmountable. How do we measure national prestige? Without dwelling too much on this obstacle, Weinberg suggests that 'among the most attractive social values that science can help to achieve is international understanding and cooperation'.[13] The most expensive scientific enterprises (accelerators, space exploration) happen to be precisely those which, by their gigantic scale, call for cooperation among nations.

The value of Weinberg's proposed criteria is that they suggest that, between the autonomy of science and its internal exigences, and the attention society pays to it in the light of its practical applications, it is feasible to define orientations which, to some extent, represent a third way—a solution which, as Edward Shils emphasises, is partly negative, since it consists of the renunciation of the extreme positions which once claimed universal validity.[14] But the shades of qualification thus introduced into the debate do not of themselves lead to rules of action which can be used as a 'code of good behaviour' between knowledge and power. Among the external criteria, only the first, technical merit, offers a formula which can be applied without ambiguity; this is because the objective to which scientific effort must be directed is, in most cases, predetermined. Within these limits, choice comes down to fixing the means connected with non-oriented research which can ensure the achievement of the desired result. By the very fact of being programmed, fundamental research is in fact oriented; if we want an atom bomb we clearly have to solve the theoretical problem of calculating the critical mass of uranium, or for the hydrogen bomb, the transition from the fission stage to the fusion stage. Weinberg is so well aware of this that he speaks of 'progammatic basic research'.

The second criterion is more original, both in its formulation and in its implications, in so far as it postulates the unity of science as a principle of the rules to be applied in deciding the merits of a given field: the more research contributes towards unifying the ideal structure of science, the more worthy it is of support; the more 'baroque' it is, and the less its weight in neighbouring fields, the less its claim for support. But this postulate of the unity of science cannot suffice on its own to exclude 'baroque' disciplines or fields. Research may be deemed 'baroque' which represents in some given field a work of 'aesthetic' refinement, unconcerned with influencing the course of neighbouring disciplines, but it may also be deemed 'baroque' in comparison with the

13. *Ibid.*, 28.   14. E. Shils, introduction, *ibid.*, xi.

'classicism' of a given field unless it immediately discloses its revolutionary content; in science, as in art, the baroque is an outlook and not a state. The memorials presented by Galois to the Institut de France, first mislaid and then rejected, might have been taken for aesthetic exercises; Galois himself was so conscious of this that he wrote that people compared him 'to those tireless men who every year find a new way of squaring the circle'.[15] And if one thinks of spontaneous generation, Pasteur's work might have been called 'baroque' by Pouchet, who looked on it as a valueless effort to affirm a theory opposed to his own, rather than as the empirical desire to refute the fantasies of his own theory.[16]

But let us go ahead and accept the criteria proposed by Weinberg without troubling ourselves about the difficulty, for the second of them, of measuring *how much* a given scientific sector contributes to the progress of neighbouring disciplines, and without asking, for the third of them, *whence and how* social merit receives its definition. Whatever their internal contradictions, the important thing is to combine them so as to make the conditions for deciding the orientation of research, if not more rational, at least more explicit. The interplay of the two sets of criteria should, in practice, make it possible to choose between fields which are in no way compatible, and of which Weinberg gives the following examples: molecular biology, high energy physics, nuclear energy, behavioural sciences, manned space exploration. The first two fields are, by definition, fundamental sciences, the third is applied science and the fourth is a mixture of fundamental research and applied research, while manned space exploration, says Weinberg, is a field which has not yet proved itself to be more than 'quasi-scientific'. Let us see how well these different research sectors do in their 'science policy' examination.

Molecular biology clearly satisfies all the criteria, the internal criteria

15. *Mémoires d'analyse*, Preface (September 1831), cited in the essay by André Dalmas, *Evariste Galois*, Fasquelle, Paris (1956), 128. Similarly, the letter of 31 March 1931 to the President of the Institute: 'You will see, Mr President, that my research has so far met much the same fate as that of the circle-squarers'. (*Ibid.*, 110.)

16. See François Dagognet, *Méthodes et Doctrine dans l'oeuvre de Pasteur*, P.U.F., Paris (1967), which clearly sets the context of Pasteur's 'strategy', particularly chapter 4 and the following quotation from Pasteur: 'You will note that I do not profess to establish that spontaneous generation never exists. In matters of this kind, you can never prove the negative. But I do profess to demonstrate strictly that in every experiment which has been thought to show the existence of spontaneous generation, observation has been the victim of illusion. . . .' (143.)

of readiness of the field and competence of the scientists engaged in it, the external criteria of importance for other biological disciplines, with very high marks for social merit, and also no doubt for technical merit in view of its possible consequences for medicine. In any event, says Weinberg, higher marks from this point of view than taxonomy or topology.

For high energy physics, excellent by the internal criteria, not so good by the external criteria; little influence on neighbouring disciplines, practically no influence on human well-being or technical development. This rather poor marking would not matter if this were not a very costly field; the only way of offsetting this disadvantage in social merit is to use the field as a medium for international cooperation.

Nuclear energy, being largely an applied effort, rises high in the categories of technical and social merit, so that Weinberg believes it deserves strong support even if it gets very low marks for scientific merit (and even, he seems to suggest, for competence—'suffice it to say that in my opinion the scientific workers in the field of nuclear energy are good').[17]

The behavioural sciences (psychology, sociology, anthropology, economics) are perfect from the point of view of scientific merit (the sciences are significantly related to each other) and of social merit (they are deeply germane to every aspect of human existence). They therefore deserve strong public support. On the other hand, says Weinberg,

> It is not clear to me that the behavioural scientists, on the whole, see clearly how to attack the important problems of their sciences. Fortunately, the total sum involved in behavioural science research is now relatively tiny—as it well must be when what are lacking are deeply fruitful and generally accepted points of departure.[18]

Finally, manned space exploration: excellent from the point of view of competent and dedicated personnel; somewhat unclear as to ripeness for exploitation, in view of the uncertainty of the human being's tolerance of the space environment (Weinberg's first article was published in 1963). By the external criteria, the marks become poor, like those of high energy physics, which nevertheless has the advantage of greater scientific validity, if less technical merit. Weinberg, moreover, does not mention this last point, except indirectly, when he suggests that in view of the quasi-scientific nature of the space adventure it would be better to devote the resources assigned to it to building roads and to schools or to anti-atomic civil defence programmes.

17. 'Criteria for Scientific Development', *op. cit.*, 31.   18. *Ibid.*

The science budget

When science policy literature refers to the criteria worked out by Weinberg, it is always in such condensed form that their self-consistency seems to be taken for granted. And yet, if one follows, step by step, the course of the argument and the examples cited, one realises to what extent the consistency aimed at is apparent only. A physicist, Director of the Oak Ridge National Laboratory, and a former member of the President's Science Advisory Committee, Weinberg is one of the research administrators most fully conscious of the social function of science and the responsibility of the scientist to society;[19] the ground on which he rests his argument is that of the social utility of science, including fundamental research, rather than its intellectual validity. In seeking to define principles of action which, in the allocation of resources and the setting of priorities, satisfy the requirements both of society and of science, he tends, if not to assess the requirements of science by reference to those of society, at least to show that there are fields of research which are more deserving of interest and support than others, for reasons which can be quite rationally formulated. But although he professes to have arrived at his criteria objectively, one cannot help thinking that, as an interested party, he proves himself, to say the least of it, not entirely impartial in judging the merit of fields outside his own (for example, in his comparison of topology with molecular biology, or his comments on the behavioural sciences, high energy physics or space exploration).

The idea that, in order to deserve substantial support, a given field must obtain 'good marks' when judged by several criteria or sets of criteria rather than by a single criterion seems to furnish a theoretically manageable scale of values. In practice, however, while the combination of different criteria seems to avoid an ideological clash between the idea of the autonomy of science and its social function, it does not thereby help to determine a rational relationship between the different possible options. The fact is that no 'major' decision involving large investments in research, even if taken on impulse, is ever single-minded in intention; above all, its implications are never confined to one single plane, scientific, technical or social. The special characteristic of modern

19. Thus, he concludes his essay with these words: 'At the present time, with our society faced with so much unfinished and very pressing business, science can hardly be considered its major business. For scientists as a class to imply that science can, at this stage in human development, be made the main business of humanity is irresponsible. . . .' (*Ibid.* 33.)

science is that it no longer commits itself alone, but spills its influence over into other fields: space exploration is quasi-scientific compared with molecular biology, but it sustains and stimulates the research system to a greater extent than molecular biology; topology is perhaps less deserving of support than molecular biology, but molecular biology may lead to interventions in human genetic material whose social consequences are unforeseeable, while topology may point the way to theoretical and practical progress in other fields of science; high energy physics and space exploration are 'passed fit' for international cooperation, but there is no reason why governments should look upon cooperation as an end in itself; the relative neglect of the behavioural sciences may perhaps be due to their lack of scientific maturity, but the possibility cannot be ruled out that with more systematic support, recognition and organization they would contribute to a better understanding and therefore to a greater mastery of the adverse effects of technical progress.

In the first place, the criteria proposed by Weinberg are impracticable and 'non-operational' when it comes to the practical issue of matching their presumed scale of values with budget priorities. Secondly, and above all, they beg the question of their own validity as criteria: *who determines the scale of values, and how?* Now, just as no man can tell in advance whether any given fundamental discovery will have major technical or social consequences, so no man can tell in advance whether the success of a technical enterprise deemed to be 'non-social' or 'non-scientific' may not have a powerful influence in fields which are, for the time being, regarded as unrelated to it. The ground of social utility on which issue is joined brings into play from the outset a scale of values which is external to scientific thought, and subjects scientific thought, however self-consistent it may be, to choices between which it is in no position to distinguish.

This limitation becomes even clearer when, in a second article in *Minerva*, Weinberg turns to the criteria which society might apply, no longer to choose between different lines of research, but to determine the total budget for science compared with other activities such as education, social security, assistance to the developing countries, and so forth. The idea of an overall science budget, he points out, is misleading; it is better for pure science and applied science not to compete with each other in the race for appropriations. A clear distinction should be drawn between the resources allocated to fundamental research and to oriented research, fundamental research being dealt with on its own, since it has not the political justification enjoyed by other forms of research by virtue of their promise of short-term applications. In this way it will be easier to assess whether the particular goal of applied research can be better achieved by scientific research than by some other means. For example, suppose we want to control the growth of population in India and we have a certain sum at our disposal for this purpose. Several alternatives are open, from the support of research activities to the purchase and distribution of contraceptives or even incentive payments to Indian women to encourage them to practise contraception. Research should be supported, says Weinberg, only as 'scientific overheads' of the process designed to make birth control more effective. Thus,

> The scientific work that goes toward solving this problem ought to compete for money with alternative, non-scientific means of controlling the growth of population in India rather than with the study of, say, the genetic code.[20]

In a sense this reasoning seems applicable to the United States system, under which not only do government agencies support 'oriented' fundamental research related to their own functions (as technical ministries or agencies do in every country), but Congress grants and discusses this support under budget procedures which deal with each case individually and independently of other agencies. In the United States scientific affairs are not presented complete in the form of a 'national science budget', and there is not even a comprehensive debate in Congress on the budget requests of the different agencies as a whole. In theory, therefore, it seems more rational for each agency to present its oriented research activities as part, not of the science budget, but of the budget allocated to the programmes in which it participates, and to regard as part of the science budget only that part of its activities which does not expressly relate to the discharge of its functions.

But not only is this reasoning, in the American context no less than in any other, purely academic, but it also masks the true nature of the research process. Pure science can be divorced from other forms of research only by statistical artifice. The research system does not *begin* at the stage of pure science or *end* with applied research; as it operates today in industrial societies, it forms *a continuous process* in which, from one stage to the next, the abilities and functions of researchers do not basically differ, the nature of research activities being no more

20. 'Criteria for Scientific Choice, II: The Two Cultures', *ibid.*, 82.

ascribed once and for all to a given category than the mission assigned to them. *Support for pure science cannot be dissociated from support for research oriented to its possible applications.* There is in fact no rivalry between the pure science budget and the budget for other forms of research, since this budget is at the same time the condition and the consequence of the aggregate resources available to *the whole research system*.

While the scientific value of the lunar exploration programme, or at least the manifest disproportion between its cost and the hope of any scientifically valuable experiments or discoveries, is open to challenge, no doubt not without reason, it can nevertheless not be alleged that this option was made at the cost of science. In the battle for appropriations it is not pure science and space technology which are in competition, but research and development activities and *all other activities*. Not that the concentration of appropriations and scientific manpower in certain sectors does not lead to major distortions. In the United States these distortions are particularly evident, first between States, the eastern seaboard being more favoured in federal grants and contracts than the middle west, to such an extent that the Americans speak freely of a 'technological gap' between the States of the Union similar to that denounced by Europeans between the United States and Europe; between universities, too, some 20 per cent of them receiving more than 60 per cent of federal research appropriations, out of a total of more than two thousand higher education institutions (of which number 120 universities alone are responsible for practically the whole national research effort); between university departments, some of them being neglected or sacrificed for the benefit of the subjects more likely to attract the public manna; finally, between the functions of teaching and research, many teachers and students turning away from teaching to give their full time to research.[21]

But it cannot be concluded from these distortions, grave as they are for the equilibrium of university development or the economic development of the different regions, that a more even distribution would have served the interests of science any better, or that other options would have been more rational from the sole point of view of the research system. One American scientist, among many others, has listed a series of alternative uses for the thirty billion dollars at which he estimated (or in spite of everything somewhat overestimated) the cost of the journey

21. See *Reviews of National Science Policy: United States*, OECD, Paris (1968), 275f.

to the moon;[22] are they more rational than the lunar programme? There is in fact no guarantee that, if the lunar programme were abandoned, the budget appropriations thus released would be used in the ways proposed, or indeed in any other way, since it is by no means certain that these resources would ever have been raised without the spur of Russian competition; conversely, there is no reason to think that the space adventure will end in the red, even in the light of other possible options. Society could, of course, allocate to other ends the resources which it appropriates to the research system, but it is by no means certain that other priorities would render the same service to science.

In most countries unwritten usage requires that appropriations to fundamental research shall be not less than ten per cent of the total research and development effort. But there is no warrant for seeking a fixed ratio between the amount of resources to which fundamental research can aspire and those devoted to other research activities. The needs of university research can no doubt be calculated from projections showing the growth in the student body and the teaching force, and it can be suggested, as Harvey Brooks has done, that a minimum of 15 per cent annual increase in the aggregate support for university science should be enough to cope with the expansion of university institutions as it can be approximately foreseen on the basis of existing structures. But even this estimate is still very uncertain, since it takes no account either of the introduction of new technologies into the pure sciences or of the changes which these technologies may involve in the pattern of industrial demand for qualified manpower.[23]

From comparison of national statistics it can further be concluded that the proportion of total research resources allocated to fundamental research is not without significance from the point of view of economic

22. 'We could give a 10 per cent raise in salary, over a 10 year period, to every teacher in the United States, from kindergarten through universities (about $9·8 billion); give $10 million each to 200 of the best smaller colleges ($2 billion); finance 7-year fellowships (freshman through PhD) at $4,000 per person per year for 50,000 new scientists and engineers ($1·4 billion); contribute $200 million each toward the creation of 10 new medical schools ($2 billion); build and largely endow complete universities with medical, engineering and agricultural faculties for all 53 of the nations which have been added to the United Nations since its original founding ($13·2 billion); create 3 more permanent Rockefeller Foundations ($1·5 billion); and still have $100 million left over to popularise science.' (Warren Weaver, 'What a Moon Ticket will buy?', *Saturday Review* (4 August 1962), no. 45, 38.)

23. H. Brooks, 'The Future Growth of Academic Research: Criteria and Needs', in *Science Policy and the University* (ed. Harold Orlans), The Brookings Institution, Washington (1968), 71.

development and, particularly, of the country's capacity for innovation. The proportion is greatest in the developing countries and the small industrialised countries (between 21 and 31 per cent), whereas in the United States and in the big industrialised countries it varies between 12 and 18 per cent.[24] In other words, one must not go too far; the greater the proportion of research activities absorbed by fundamental research, the less the research system seems to be integrated in production as a source of applications to be exploited. The countries with proportionally the largest budget for university research clearly seem to be those whose industrial structures lend themselves least to technical innovation. But what is the proper proportion?

The needs of fundamental research depend *primarily* on the skills available and the fields opened up by the unsolved (or unformulated) problems of science itself. The needs of applied research, on the other hand, depend primarily on the problems which the industrial system desires to solve. There is no hermetic seal between the first type of problem and the second, the terms of each being renewed or changed by the progress made by the other on the basis of a certain degree of osmosis between the university system and the industrial system. *But if society can at a pinch measure the price it attaches to solving the second type of problem, neither society, nor indeed the scientists themselves, can measure the price they attach to solving the first type: the ideal volume of the pure science budget cannot be determined by any rational calculation.*

It is, moreover, striking to note that the custom of devoting about one tenth of the total research and development effort to pure science was advocated long before the days of industrialisation and affluence. Condorcet, in his *Fragment sur l'Atlantide*, had already made the use of the resources available to the 'society of savants' subject to the condition that

> One tenth of the subscription, let us say, shall always be set aside to serve the general views of the association, in order to ensure that its utility extends to the whole system of human knowledge.[25]

The criteria of choice proposed by Weinberg no doubt make the issue of the debate more explicit, but they do not lessen the ambiguity of the debate itself. By accounting for part of the programme or activities of a research agency under the title of 'scientific overheads', Weinberg pro-

24. See *The Overall Level and Structure of R & D efforts in OECD Member countries*, OECD, Paris (1967), 34.
25. Condorcet, *Oeuvres complètes*, Garat-Cabanis, Paris (1804), 595.

poses a frame of thought in which the value of all scientific activity is
assessed by the standard not of its value to science alone, but of the
solutions, direct or indirect, which it provides to the problems of society.
Applied research is treated as the 'overheads' of the technical programmes with which it is integrated, oriented fundamental research as
the 'overheads' of the applied research to which it contributes, and,
finally, pure research itself is defined as 'overheads' of activities supported by the public authorities in the context of technical progress.[26]
This cumulation of 'overheads' does not solve the problem of setting
priorities between disciplines or fields of science or, especially, between
research activities and other activities, but it clearly confirms, from one
stage to another, that science as such cannot rely only upon itself to 'deserve' the public support it demands.

Strategies and models

In the debate opened by *Minerva*, Weinberg's contribution lies halfway
between two concepts: Polanyi's vision of a republic of science in which
things go all the better the more they are left in the hands of scientists,
the spontaneous trade unionism of researchers leading to more rational
choices than could ever be determined by the most coherent effort to
orient science from the outside; and the resolutely 'integrationist' concept of an economist such as C. F. Carter who refuses any kind of
autonomy to scientific research, even non-oriented, and wants to determine its directions in the light of economic needs as visualised by
long-term planning—the pattern of scientific effort should be designed
with the maximum efficiency to increase the flow of national wealth.[27]

Taking the example of the United Kingdom, Carter stresses that the
most important long-term economic need is to increase exports:

> A starting-point for scientific policy, therefore, is the examination of what
> we know about the pattern of exports ten or twenty years ahead. . . . By
> asking these questions about long-term prospects of British exports (and
> about substitution for imports) it would, I think, be possible to reach
> some conclusions, first about applied research and development, then about
> the pure research which feeds into it, and finally, about the forms of
> training needed to support this research.[28]

This theme is in fact that of all the industrialised countries which are
compelled by international competition to make their presence felt in

26. 'Criteria for Scientific Development', *op. cit.*, 90.
27. C. F. Carter, 'The Distribution of Scientific Effort', *ibid.*, 34–43.
28. *Ibid.*, 40–1.

technical innovation; it assumes that research strategies take second place to production strategies.

Economic imperatives allow little subtlety in the function which must then be discharged by research: where Weinberg still left some room, however narrow, for scientific considerations, Carter dismisses all argument based on the specific nature and interests of science. The total pure research effort must be considered in two parts:

> One part would be capable of justification by its ultimate usefulness in application, *as well as* by its inherent interest; the other part would be a work of supererogation, justifiable only by the pure joy of discovery. A nation which chose to undertake pure research beyond what can be given a broad economic justification by its ultimate application would have regarded intellectual discovery as more important than an increment of wealth.[29]

This is not exactly the case with Britain, adds Carter, since it cannot be proved that the total scale of pure research exceeds what can be economically justified; there is nevertheless a disproportion between pure science and applied research, no doubt because pure science enjoys greater social prestige in Britain.

It may be noted that this way of formulating the problem does not lead to clearer or more manageable formulæ than the others, even if Carter, by an act of faith in the virtues of planning, does not despair of this result:

> There will be a temptation to suppose that the problem of the distribution of scientific effort is too difficult, complex or vague to be solved. Yet a few tentative moves towards rationality may be better than leaving the disposal of a valuable resource to random and undirected influences. One hopes that this is why we have in Britain a Minister for Science.[30]

Summary as it is, this concept has the apparent advantage of avoiding the ambiguous reasoning of Weinberg, which justifies support for pure science under the banner of considerations, programmes or activities which are external to it. In Carter's eyes, since society's reasons for supporting pure research are extra-economic, the problem of choice comes down to the choice between applied research and technical development; the overspill of research which seems to have no economic justification must be entered as 'profit and loss' rather than as 'overheads'. A country is free to invest in research for the pure pleasure of the researchers, but not only are there more pressing priorities, there is also a more rational road to be followed if science is to meet these priorities.

29. *Ibid.*, 37.   30. *Ibid.*, 43.

Now to identify this road in fact is nevertheless the question which remains in suspense, since neither theory nor practice provides a very helpful answer. On the practical side there are no doubt strategies which are more effective than others—one thinks of Japan—in harnessing a country's scientific effort to its economic development and international competitiveness; but these strategies are empirical and not the application of formalised models which can be transposed into contexts other than those in which they have been empirically worked out. Even so, we must not be under too many illusions about the role of governments in orienting research efforts in the private sector. The Japanese example is indeed a case apart, both because of the close and confidential relations—informal but highly efficient—which exist between the political decision-makers and the heads of big business, and because of the control which the government has always exercised over industrial investments; the essence of these relations rests more on national consensus than on exemplary procedure for concerted economic action.[31]

In most market-economy countries the situation is very different, even in France where the planning mechanism is nevertheless reputed to assist in concerting the investments of the public sector with those of the private sector. Not only is French planning purely indicative, but it is also common knowledge that government intentions are far from carrying great weight with the leaders of industry, especially in the matter of research: in the first place, because the idea of a scientific strategy linked to the strategy of production is recent, and French business firms are traditionally disinclined to take the financial risk of big research programmes; and secondly, and above all, because the terms in which such a strategy can be formulated are all the vaguer since the government's priority options leave little margin of resources for research activities not directly connected with those options.

Nothing is more revealing in this respect than the report of the expert group on 'Research and the Economy' which was convened in 1967 by the then minister in charge of scientific research. This report distinguishes three types of research affecting industry: 'big operations', such as the generation of electricity by nuclear power, the space

31. See, in particular, in *Reviews of National Science Policy: Japan*, OECD, Paris (1967), the Examiners' Report: 'There appear to be a number of informal methods by which government and industry cooperate and influence one another . . . quite apart from the institutional structure of research' (17), and this comment by Professor Piganiol, one of the examiners: 'Everything happens as if, beneath the structures and the organisation charts, the power of persuasion was accompanied by a very powerful expression of the national will.' (56.)

telecommunications programme, the computer plan or the Concorde supersonic aircraft; 'middle topics', oriented towards technical innovation programmes; and 'diffused research', the support of which would enable a branch or a firm to retain or acquire competitive capacity.[32] On the one hand, these last two categories represent less than 15 per cent of the resources envisaged as government support for research in the private sector; this means that support for research unconnected with 'big operations' is confined to scattered actions, chosen more in the light of casual circumstances than by deliberate policy. Secondly, hardly has the expert group affirmed that they must be subject to economic criteria, when one reads that they should be chosen

> by comparing them with each other and applying to them, from the technical and economic point of view, qualitative criteria based on their contribution to general industrial progress.[33]

This transition from the quantitative to the qualitative is a delicate euphemism for treating decisions of an eminently political character under the cloak of economic rationality; the qualitative criteria are those by which the government deems, rightly or wrongly, that it is in its interest to support some particular orientation of research regardless of considerations of economic return.[34]

Furthermore, even if the economic choice were as simple as Carter thinks, it is not enough to influence *one only* of the elements of the research system to ensure that science policy has most of the luck on its side. In the light of the government's economic aims, it may seem more logical according to Carter's line of thought to have, as formerly in Britain, a Ministry of Technology separate from the Ministry of Science or, as in France, a Ministry of Industry and Scientific Development succeeding the minister in charge of scientific research. If science policy oscillates between an impossible rationality and neces-

32. Report of the Expert Group on Research and the Economy, Office of the Commissioner-General for the Equipment and Productivity Plan, in *Le Progrès scientifique*, Paris, DGRST, no. 119 (May 1968), 7–8.
33. *Ibid.*, 7.
34. 'Calculations of economic return are, in practice, inadequate for assessing projects of this kind, whose most important economic effects are often indirect and diffused. The real criteria should be the effect of the project in stimulating and bringing closer together scientific disciplines and branches of industry capable of profiting from it, the clearly demonstrated need to make up a European lag behind the United States, the possibility of filling a niche left vacant by American technology, and the possibility of encouraging the creation of a large business on the basis of the initial operation' (*ibid.*). Each of these criteria seems to find economic justification after the event for 'major projects' which have in any event already been launched.

sary strategies, the policy of orienting the whole research effort in the light of its utility to the production system is certainly the most tempting. But this logic is valid only up to a certain point, since it introduces between the different elements of the research system the very frontiers it professes to suppress, confirming university research in its 'ivory tower' by turning it into a ghetto. The minister of science then ceases to appear as the spokesman of science with the powers that be and is no more than the representative of power *vis-à-vis* science; we can hardly be surprised if, in reliance on long-term economic considerations, he is likely to pursue a short-sighted policy. By denying the ambiguity of the status of research in the production system, we do not escape the need to ask how and in what conditions free research must be undertaken if the priority decisions on applied research are to be crowned with success.[35]

On the theoretical side, there is no lack of attempts to construct strategic models. But the methodological or heuristic interest of these models does not go as far as making them applicable to decisions affecting science. The more the specialists try to integrate the production and diffusion of technical knowledge with economic analysis, the more they find that the factors they have to take into account to arrive at satisfactory models are multiple and heterogeneous. These efforts are something like a tower of Babel whose base will never be large enough for its summit to touch the sky; a plethora of factors to be processed, and the process uncertain. If there is a lesson to be drawn from the work of V. V. Leontieff on calculating an 'index of structural change', it is clearly that the decisions on which that change depends are so numerous and of such different relative weight that their comparison in the most elaborate of tables will never provide more than a very artificial cross-section of the process.[36] At the level of government action,

35. It might be thought that it was easier to allocate research appropriations 'scientifically' at the micro-economic level of business firms. In practice, even at this level, methods of decision are not necessarily more elaborate or motives more resistant to extra-economic considerations, such as prestige or the habit of maintaining the volume of work or of personnel at a constant level; owing to the steady increase in the cost of these items many firms habitually increase their research budgets by at least 5 to 10 per cent each year for no other reason than that. See, in particular, 'Research appropriations in United States firms are not allocated rationally', report of an address by Albert H. Rubenstein, Professor of Research Administration, Northwestern University, in *Le Monde*, Economic Supplement, iv (22 July 1969).

36. See V. V. Leontieff, *Studies in the Structure of the American Economy*, Oxford University Press, New York (1953), especially the conclusion of chapter 2: 'As the elements of the economic system change and become finally

in any event, the assurance that a strategy inspired by these models is the sound one, is as limited as a medical diagnosis based solely on microscopic examination.

A particularly telling example of this limitation is provided by the attempt of Claude Maestre to construct an overall model of exchanges between research and production.[37] On the basis of Leontieff's input–output tables, Maestre proposes to calculate the effects of research on the different branches by defining them in terms of national accounts. His model consists of a series of grids in which the scientific fields, represented by numerical indices, are related with the branches of production to which they may contribute. There is no point in listing here all the methodological difficulties of this attempt which, in the author's mind, 'opens the way towards an instrument of a veritable strategy of research linked to a strategy of production'.[38] For example, in order to calculate the research operations necessary for a branch, the values assigned to these operations must be translated into quantitative representations; but these values can be assigned only in the light of subjective criteria of utility, and presuppose a relation of a mechanical type between the investments to be made in any given discipline and the increase in products due to research in the branches concerned. More significant is the difficulty, if not the impossibility, of introducing into the model the dynamic dimension of time; for decisions affecting a process whose results are not obvious in the short term, what can be the value of a 'snapshot' of this process, however meticulous? The instrument of the strategy of research proves to be scarcely even a General Staff map.[39]

_____

unidentifiable in historical terms, the scope for the application of analytical and descriptive devices based on simple or even sophisticated numerical comparisons is slowly but inexorably narrowed down. The investigator who refuses to give them up in favour of more general methods of quantitative analysis will find himself finally completely enmeshed in a maze of unintelligible time series and heterogeneous aggregates'. (43.)

37. C. Maestre, 'Vers une mesure des échanges intersectoriels entre la Recherche et l'Industrie' in *Le Progrès scientifique*, Paris, DGRST, no. 102 (November 1966), 244.

38. *Ibid.*, 6.

39. As, indeed, the author has recognised in a clarification published subsequently: 'The decisional, and even the operational, value of the method proposed must be carefully qualified. As things are at present, we must, indeed, refrain from any use which, through this representation, would postulate the existence of formal relations between research and industry'. (*Le Progrès scientifique*, Paris, DGRST, no. 119 (May 1968), 47.)

Research and utilitarianism

Let us halt at this point; it is clear that, owing to the nature of research activities, the choice among them at government level cannot represent a rational calculation in the sense of expressing the logic of scientific thought. The terms of the relation between science and power remain conditioned by the objectives of power rather than by a rationally established conjunction between the objectives of the one and the objectives of the other; the *scientific* debate on the overall orientation of the research effort is doomed to be one *political* debate among many others.

Since this relation cannot be formalised in the language and terms of science, it may well be asked how it happens that so many intelligent people, scientists, economists and administrators, nevertheless believe that it can and dedicate themselves to proving it; what, in sum, is the source of the mythical hope of objective and rational criteria of choice for the allocation of resources to science? It is not enough to answer that the idea of subjective factors affecting decisions, scarcely made explicit by the methodical approach and inconclusive in terms of strict logic, is alien or intolerable to the scientific mind; or that science, called upon to define the best means of achieving certain ends, deludes itself about the capabilities of its own procedures by refusing to recognise that the definition of means is one and indivisible with the choice of ends.

Quite evidently this positivism makes itself felt in the debate on the allocation of resources to science, as though the rules of scientific thought had only to be recited in order to install on the political stage a décor as rational as their own; if the research effort can and should be more rationally oriented, it is on condition that it is freed from the pressures of ideologies, values and conflicts of interest which constitute the inevitable texture of political decision; once this texture is supplanted by scientific reasoning, the methodical consideration of means should sweep away all the disadvantages of the partial, ambiguous and not universally convincing political consideration of ends. But this positivism is still only a sign or, if you like, a behaviour, and not an explanation of this behaviour. The naïvety with which some researchers expect political procedures to conform to the rules of scientific procedures does not explain why they do not despair of transforming into a rational operation what can at best be no more than the subject of a rough reasoned calculation.

In spite of the fact that the extension of knowledge for its own sake is only one element among others in the research system, it refers back to values which are never based on the criterion of utility alone. Judged by the eyes of truth, science is a 'cultural good'; judged by the eyes of utility, it is a force of production. On the one side it is a value in its own right; on the other it is an exchange value only, measured by its results. The debate aroused by this ambiguous status recalls the debate which raged around utilitarianism in the fields of ethics and law; when applied to the relations between science and power, utilitarianism comes up against the same impossibility of basing the values it advocates on empirical facts or concepts. It can be shown that a means is valid for achieving an end already set, but the validity of the ends themselves cannot be directly proved.[40] The ends themselves must be accepted as a datum; they are what men in fact desire and seek.[41] It follows that the theory of ends derives from experience itself, whether it relates to psychology, as with John Stuart Mill, or to any other social fact. This same process is at work in the efforts of scientists to formalise the terms of a contract between science and power; the rule of decision for the orientation of the research effort is therefore determined by social utility and, in this case, empirical teleology is founded on the economy. But just as psychology, for example, provides no scientifically valid principles by referring the motives of all acts to personal interest, so the economy provides no strict principles by referring all the motives of the research system to the general interest.

Utilitarianism is condemned to attach greater value to certain interests than to others, by justifying them by their utility after the event; a 'maid of all work' principle, in sum, which assumes that the reasoned search for the desirable in fact coincides with the general interest. The difficulty begins when it comes to distinguishing among desirable things, some of which have a greater qualitative attraction than others; the only way out then is to count heads of those who prefer the one or the other.[42] Similarly, one must assume a happy conjunction between the deter-

40. For example, John Stuart Mill: 'It is evident that this cannot be proof in the ordinary and popular meaning of the term. Questions of ultimate ends are not amenable to direct proof. Whatever can be proved to be good, must be so by being shown to be a means to something admitted to be good without proof' (*Utilitarianism*, Everyman's Library (1971), chapter 1, 4).

41. 'Questions about ends are, in other words, questions what things are desirable. The utilitarian doctrine is, that happiness is desirable, and the only thing desirable, as an end; all other things being only desirable as means to that end.' (*Ibid.*, chapter 4, 32.)

42. 'Of two pleasures, if there be one to which all or almost all who have experience of both give a decided preference, irrespective of any feeling of moral obligation to prefer it, that is the more desirable pleasure.' (*Ibid.*, chapter 7, 8.)

mination of the utility of science and the determination of social needs if the choice of the orientation of the research effort is to be an indication of the general interest. *The insuperable obstacle to science policies is the impossibility of measuring the utility of science*; the criterion of utility amounts in this case, as in the case of ethics, to the validation after the event of choices which reflect preferences rather than referring to a universally applicable scale of values.

Furthermore, the legitimacy of the ends is not the only issue; there is also the relation between means and ends. The logic of decisions affecting science would be fully rational if calculations of economic return, models of maximisation and optimisation based both on a cost estimate and on an assessment of the chances of success, permitted input–output relations tainted with a minimum of random error. Not only is this not the case, but we cannot content ourselves with imputing this limitation to the defects of economic techniques without postulating, to the point of absurdity, the minimisation of the uncertainty inherent in all research activities.[43]

This becomes specially evident in the efforts of certain economists to calculate the part played by research activities in economic growth. It is self-evident that economic growth is constantly more dependent on technical innovation and that innovation largely depends on science. But exactly how big is 'largely'? (and what, moreover, is the exact role of technical innovation in economic growth?) The answer is that no reliable assessment can be made—at any rate so far. Factors as remote from science proper as management techniques, market organisation, manpower mobility, natural productivity growth and so forth, seem to be no less important than scientific discovery or technical innovation. The concept of the 'residual factor'—everything which contributes to the growth of production without changing the quantity of input, education, development of knowledge and skills, scientific research, etc.— marks a certain progress in economic theory, since it recognises the role of changes in the social and technical context in which prices and in-

43. See F. Machlup. Inquiring into the rationality of federal R & D programmes in the United States, he leaves no room for ambiguity: 'The choice of major research priorities is largely a political one. There are no simple cost-benefit criteria, for instance, which will decide whether health R and D expenditures should be increased at the expense of military R and D expenditures. The major choices are largely value judgments. . . . Political pressures, the activities of special-interest groups, as well as historical accident, also play a role. It is difficult to see how this could be otherwise, since many variables that are of crucial economic importance cannot be measured very accurately.' (*The Economics of Technological Change*, Norton and Co., New York (1968), 187.)

comes fluctuate. While it is clear, however, that residual factors have an important influence on economic growth, it is impossible today to specify the exact relative importance of each of the factors which come into play, and everything indicates that it will always be impossible; as has often been said, the residual factor is, first and foremost, a measure of our ignorance.

According to Denison's study, the fullest on the subject, only about one-fifth of the 'advancement of knowledge' factor, said to have contributed about 30 per cent to the economic growth of the United States between 1929 and 1959, can be attributed to organised research and development.[44] It is true that Denison's calculations do not take account of improvements in the quality of products, to which a great part of research and development work is in fact devoted. Furthermore, the considerable growth in research expenditure since 1957, as well as the measures taken in the United States to organise the structures of science policy, have changed the data on which he based his first calculations. For the period 1950 to 1962 he now estimates that the proportion of the 'advancement of knowledge' factor has increased by about 0.1 percentage points, which he regards as a substantial change for the particularly efficient and diversified American economy.[45] But while recognising this change as an effect of the new scale of support for research activities, Denison again emphasises that it is impossible to base any calculation whatever of the effect of such activities on the economic growth rate, on the amount of money invested in them. Quite evidently, the process is too complex for any relation of cause and effect ever to be strictly established between the intensity of the research effort and the rate of evolution of productivity or expansion.[46]

44. Edward F. Denison, *The Sources of Economic Growth in the United States and the Alternatives before us*, Committee for Economic Development, New York (1962).

45. Edward F. Denison and J. P. Poullier, *Why Growth Rates Differ: Postwar Experience in Nine Western Countries*, The Brookings Institution, Washington (1967), 282.

46. 'Figures such as these are impressive and suffice to show that the United States was devoting more effort than the European countries to R & D. Their relevance to an analysis of measured growth is remote, however. They refer to inputs into research and tell nothing of what is learned. The purpose of most— probably the bulk—of R & D expenditures, moreover, is such that it does not affect the growth rate no matter how successful it may be. The growth rate would have been the same . . . whether antibiotics were developed or not. It would have been the same if the effort to reach the moon by 1970 had not been made and, instead, the same resources had been used to build and stockpile battleships. It will be the same whether the moon is or is not reached by 1970.' (*Ibid.*, 288.)

Even if such a relation could be established, it would still have to
be explained in what conditions the overall research effort, and more
especially non-oriented research, influences economic growth. The
mythology in which science policies are immersed cannot conceal the
fact that there is no necessary relation between a country's prosperity
and the size of its research expenditure. Italy, which devotes a ludi-
crously small part of its gross national product to research and development, enjoys a rapid growth rate, while the United Kingdom, with the
largest research expenditure in Europe, is far from being in an era
of prosperity. It has been the same with France since the war; whereas
its gross national product has almost doubled in fifteen years and the
standard of living is twice as high, and in spite of a population nearly
7 million more than in 1938, research and development expenditure ten
years ago was still less than 1 per cent of gross national product. At the
very least, this indicates that, even if economic growth is taken as the
end, the utility of research as a means is not thereby a decisive measure.

When Paul Valéry, with some apologies, borrowed stock exchange
language to speak of spiritual values—

> the spiritual economy, like the material economy, when one comes to think
> of it, can both very well be summed up as a simple conflict of *evaluations*

—he sought to limit the parallel by showing that spiritual values,
unlike economic values, cannot be measured by any instrument:

> The economy of the spirit presents us with phenomena which are much
> harder to define, since in general they are not measurable, nor are they
> vouched for by specialised *ad hoc* institutions or bodies.[47]

He could not foresee that the material economy itself would come up
against the same limitation in relation to the research system. While the
research system can always be treated as one system of production
among others, in which the promises of output are measured by the
volume of input, it is nevertheless impossible to infer from scientific
reasoning the criteria for a more rational distribution of resources. The
concern displayed by scientists to rationalise choices affecting science
amounts in fact to treating this system as though it were no different
from other systems of production, or, in other words, recognising that
the objective criteria which they pursue are not fully within their competence. In that, moreover, they do no more than ratify the economic
thesis of utilitarianism that the process of invention and change in

47. Paul Valéry, 'La Liberté de l'esprit', in *Regards sur le monde actuel*,
Gallimard, collection 'Idées' (1958), 265, 268.

general does not constitute an exogenous variable in the production system, still less an independent variable. It is Mill himself who writes:

> In a national, or universal point of view, the labour of the savant or speculative thinker is as much a part of production in the very narrowest sense, as that of the inventor of a practical art.[48]

In the relations between science and power, each of the partners has learned to deal with the other and to derive benefit from the other, but one of them has nevertheless found it easier to learn than the other; the scientists have had to learn to behave like politicians, whereas the state needed no lesson. The orientation of the research effort is decided on the ground which the state has chosen in the light of the challenges, real or imaginary, which face it, strategic threat or international competition, the race for power or prestige in connection with the one or the other; its reasons for influencing the growth of the national scientific effort are external to the ends of science. The measure of utility is in its hands, not only because they hold the purse strings, but also because the urgency with which a given orientation of the research effort demands priority is circumscribed by the nature of the needs it feels bound to satisfy. The fields of science and technology which contribute most to technical change are precisely those in which the nations have reason to expect an instrumental solution to the problems they set themselves. From the outset, the powers that be place their criteria on the level of utility *to themselves*; science is one of their objectives only so far as it is a means towards their ends.

On the other hand, free research regards utility only as an indirect objective, mediated by the search for truth, science for its own sake, or the simple but exalted pleasure of seeking (and finding); external criteria take their place, if not after internal criteria, at least in the same rank. But these internal criteria are not those which determine the orientation of the research effort: *the legitimacy of science conceived as a value in itself is masked by its recognition as an exchange value.* And it is only on condition of assuming this mask that science can join in the power game; the area of decisions affecting its future is in no way that of the rationality which it invokes as a consideration of truth. Scientists, like any other agent of the political process, take advantage of the favours which the powers that be accord to them when in need of them. Their definition of utility does not coincide with that of the

48. J. S. Mill, *Principles of Political Economy*, D. Appleton and Co., New York (1890), volume 1, 68, cited by J. Schmookler, *Invention and Economic Growth*, Harvard University Press, Cambridge, Mass. (1966), 208, note 2.

state, except in so far as they feel themselves recognised, and with them their special field, as one of the ends pursued by the powers; they are then only this end because they also offer themselves as its means.

The relation which unites knowledge to power is in no way a *contract*, 'an agreement between two or more persons', as defined by Littré, 'designed to create or extinguish an obligation'. The obligation of support which scientists expect from the state is conceived by them as having no counterpart. And the state itself could not impose a strictly contractual relation on researchers without discrediting them in their own eyes and in those of the public, precisely in so far as they hold themselves out as researchers. This relation is, in sum, one of bad faith, that is to say, a lie *per se*; each partner excludes from his behaviour what he expects from the other, and it is by masking their common dependence under the cover of their divergent objectives that each of them grants the other what it professes to refuse him. The researchers rely on the specific nature of their activities to escape the requirements and even the control of the state, or claim a privileged relation to evade the other demands made on the state—a 'preferential' relation, in the language of international trade, or, if you like, an alliance in the sense in which the Jewish people speak of the alliance with God. The state is not God; it has to reckon with other choices, and the duties which it renders to research remain dependent on the choice it makes to meet the needs of the one against the exclusive demands of the other.

The rule of the relation between science and power obeys not the logic or values of scientific thought, but the contingent, partisan and conflicting pressures of the political process. It is within the competence of scientists to define utility not because they have a greater mastery than other people over the measuring instruments on which it depends, but because they themeselves as citizens, on a par with their non-scientific fellows, desire to see it defined in a certain way. They can change the orientation of the research effort and make it more rational and more in line with the universal intentions of scientific thought, not by influencing the research system as a means at the service of the state, but by influencing the ends of the state themselves.

PART III

Science in Politics

If the experience of science could be communicated—and I think it important that it should be—it would prepare far more individuals for the difficult situation of 'man in face of the universe'—a situation about which philosophers and governments seem to me today to be hopelessly out of date.

OPPENHEIMER

CHAPTER 7

The Scientist and the Problems
of Power:
The Discovery of Responsibility

According to Max Weber, there are only two ways of making politics one's vocation: 'Either one lives "for" politics or one lives "off" politics.'[1] The two ways are not mutually exclusive; the quest for power as a 'life aim' spills over into the search for 'a permanent source of income', the vocation becomes a profession, since a man who wants to be politically active must be economically independent. But the savant, looking down from his professorial chair, must eschew both ways, for the very reason that, being neither prophet nor demagogue, it is not his function to impose his personal views. Weber's papers collected under the title of *Science and Politics* sharply contrast the vocation, the attitude and the profession of the scientist and the politician, since 'politics is out of place in the lecture-room' on the part either of the student or the teacher, 'and when the teacher (the "docent") is scientifically concerned with politics, it belongs there least of all'.[2]

It is in and through his function as a teacher that the scientist is distinguished from the politician: the professor is not a leader; his scale of values not only differs from that of the politician but is inconsistent with it. Furthermore,

> The qualities that make a man an excellent scholar and academic teacher are not the qualities that make him a leader to give directions in practical life or, more specifically, in politics.[3]

Naturally, Weber did not rule out the possibility of the savant's adopting a political attitude or even being drawn into a life of action by the very factors which determine his culture and objectivity as a man of learning, on condition that his action lies outside the university, in his

1. Max Weber, *Science and Politics*, 23.
2. Max Weber, *Science as a Vocation*, 145.
3. Max Weber, *Science and Politics*, 150.

capacity not as a master but as a citizen. The temptation was one which, as everyone knows, he continued to feel; in this sense, he went further than Renan in proclaiming that the service of science is suspended by political combat.[4] 'The impossibility of "scientifically" pleading for practical and interested stands' did not stop Weber from thinking that, even in political combat, science can still 'help to gain clarity'.[5] As a general rule, however, science as such was not in his eyes a partner in the political enterprise. And no doubt at the beginning of the century it was not so to the same extent that it is today.

The fact is that there is nowadays a third way of making politics one's vocation, and the scientist as such can no longer eschew it; not living for politics or off politics, but living *in* politics, as a consequence and not as a cause of one's profession, as a destiny superimposed on one's vocation, rather than the encounter in it of different dispositions. The problems raised by the administration of modern societies whose development, and even whose survival, depends on scientific knowledge, make scientists the inevitable interlocutors of power. Bound by the means which power grants them, sometimes also by the secrecy it imposes upon them, they are also bound by the ends which it pursues through the medium of their work. The state's need to mobilise scientific resources and rapidly to convert theoretical knowledge into applications automatically means that the politicians are dependent on the scientists; not only has science ceased to be irrelevant to the political enterprise, but it conditions its ways and means.

It was in the capacity of expert that the scientist first discovered this other way of following a political vocation; no more a chief, certainly, but adviser to the chief—'on tap but not on top', in the American phrase—he nevertheless exerts an influence on the administration of public affairs. This function of counsellor has become a necessary institution in the governmental machine of modern societies; but science is only one example among others of technical questions which are so complex and specialised that the political authorities cannot handle them without consulting experts. In this respect, the scientist's role is no different from any other technical function associated with public

4. 'There are, I confess, sciences which could be called shadow-loving ("*umbratile*"), which seek security and peace. It took Monsieur de Sacy to publish, in 1793, at the Imprimerie du Louvre, a work on Persian antiquities and the medals of the Sassanid Kings. . . . One can find moments for other duties; but it must be a suspension and not an abdication.' (Renan, *L'Avenir de la Science*, Calmann-Lévy, Paris (1890), 418.)

5. Max Weber, *Science as a Vocation*, 147.

affairs, and the resultant problems in the exercise of power are not specific. They are the problems of technocracy in the sense in which it reflects 'the ascent of the possessors of technical knowledge or skills at the cost of politicians of the traditional type',[6] of the possible engrossment of the political function by virtue of an influence based on technical skill, and in the extreme case of a form of government in which decisions are based mainly on technical considerations.

If the natural sciences have taken longer than the social sciences, especially economics, to gain recognition as a nursery for government advisers, it was nevertheless with them that the political function of the technician started. The whole development of science and its growing integration in the social system are the source of the influence of scientists in general, the specific characteristic of industrial societies being that all their undertakings depend on the methods and achievements of science. The industrial system which is the basis of the rise of the 'technostructure' is itself the consequence of the expansion of scientific research, that is to say, of the constantly closer link between theory and practice. Science, as achieved technology, elevates into the functions of direction the man who *realises*—in both senses of the word, in idea and in practice—the exploitation of knowledge.

In a sense this is a Marxist theme, whose social consequences Marx failed to see, because he prejudged its economic consequences; the reversal which has taken place in the relation between the production system and the system of knowledge—in which 'the appropriation by man of his own universal productive force', science and the mastery of nature, becomes 'the master pillar of production and wealth'—has not brought about the collapse of production based on exchange value, but it has brought about the rise to power of those who possess 'the materialised power of knowledge'.[7] In becoming an 'immediate productive force' science transforms not the economic function of production but its processes and, at the same stroke, the social function of producers. Saint-Simon was nearer to seeing what was implied in the dynamics of industrialisation in the scientific age. Scientists and technicians are vested with functions of direction only on the condition of discharging them according to the wishes of the industrial system itself:

> Savants render signal service to the industrial class, but it renders even more important services to them; they owe it their very *existence*. . . . And

6. J. Meynaud, *La Technocratie*, Payot, Paris (1964), 27.
7. Karl Marx, *A Contribution to the Critique of Political Economy*.

so it is entitled to say to the scientists, and even more so to all other non-industrialists, 'We are not prepared to feed you, house you and clothe you, and, in general, to satisfy your physical needs, except on such and such condition'.[8]

The subordination of the community of knowledge: the dynamics of industrialisation installs in the command posts all who contribute special knowledge, but this does not mean that technicians take over the political function in the sense that they alone set the goals and take the decisions. More precisely, if the government machine has come to depend on scientists, its goals are not identical with theirs for the sole reason that they contribute towards achieving them. Galbraith is careful to distinguish 'the educational and scientific estate' from the 'techno-structure', emphasising that

> there is potential competition and conflict between [them] *growing out of their respective relations to the state*.[9]

Living in politics, taking part in decisions, inspiring them (or challenging them), certainly does away with the ideal frontier which Weber drew between the champions of the ethic of conviction and those of the ethic of responsibility; but it does not follow that the latter divest themselves of power in favour of the former. On the contrary, it must be recognised that there is a close connection between the inevitable collusion of knowledge and power, and university unrest or student revolt; it is not easy to challenge the aims of authority when everything you are, everything you believe and everything you do becomes the objective instrument of its power.

We must be careful not to project into the present a past state of affairs; in showing itself in its political function, the concept of science is not returning to its Greek origins. Not only does the economic and technical landscape of the modern world bear no relation to that of Plato, to whom the paths of knowledge were the royal road towards solving political problems, but science itself, which contributes so largely to fashioning that landscape and implanting in it, to the point of saturation, the artifices of anti-nature, is in no way the same. The new relation between knowledge and power does not reopen the old debate on the political competence of scientists, in spite of the professions of faith of some of them to the effect that scientific culture, methods and

8. Saint-Simon, *Catéchisme des industriels*, in *Oeuvres*, Anthropos, Paris (1967), volume 5, 25.

9. J. K. Galbraith, *The New Industrial State*, Houghton Mifflin, Boston (1967), 294 (my italics).

objectivity should vest them with special authority in conducting the affairs of state. This idle debate suggests at worst the image of the researcher as the initiate of mysteries, the great sorcerer, the new priest of a religion whose esoteric knowledge is merged with a privileged power of decision; at best, the Platonic dream of the philosopher become king. Scientists are not expert in politics; their competence in their own more and more narrowly specialised field gives them no greater authority in other fields than any other technical skill. Politics has found in them deacons rather than priests, and no doubt the only feature which still links them with the Platonic vision is that nowhere and at no time, in the days of the scientist–philosopher any more than in those of the philosopher–savant, have they ever held the power themselves. The debate, in our day, does not lie there; *science has no special claims to challenge the rationality of politics, but politics challenges the rationality of the scientific institution.*

Death of the savant, apotheosis of the scientist

The injunction on the members of the Royal Society 'not to meddle with Theology, Metaphysicks, Morals, Politicks, Grammar, Rhetorick or Logick' refers to the days when the natural sciences, in order to establish themselves as sciences, aimed at being a universal realm of thought independent of the values which committed men in the commonwealth. This was the time of science militant, working to emancipate itself from the dogmas of philosophies and religions, from subjections to the church or to monarchs; the mathematical reading of the language of nature banished from its territory everything it could not discuss rigorously and objectively. But the age of science triumphant turned this process into a political process; since the theory of science cannot be separated from its practice, scientific activity affects society in its institutions and values and, in its turn, is steeped in the conflicts which divide cities or set them against each other. Between its preaching of neutrality and the reality of its commitment, science is now established at the heart of the antinomies of practical life from which it always professed to be free by reason of its objects and its methods.

It has, it is true, always been faced with these antinomies, sometimes as a threat, sometimes as a temptation, from the moment it broke with the ancient concept of contemplative knowledge, without mastery over things, indifferent to everything in experience which is a pretext for experiment and action. By conceiving nature in the light of instrumentality, it brought itself into the light of utility: knowledge defined as power had only to become a reality in order directly to influence the course of the world; even so, in order to become a reality it had to establish its identity as an autonomous institution among the institutions of culture. In the eighteenth century, once the bonds between theology and physics had been severed, the natural sciences made their appearance as the model to be followed by social practice.[10] In postulating the unity of reason, the century of the Enlightenment looked to the progress of knowledge to achieve the progress of the human race; it could not conceive of science abdicating its 'scienticity' to intervene in social practice. And yet this was the time when science began to emancipate itself from other intellectual activities and to dissuade researchers from taking part, as such, in the tumults of practical life. The specialisation and professionalisation of researchers were soon to consolidate what was already adumbrated by their membership of an intellectual community whose work and norms were controlled by them alone, namely, a technique which acted on the world with the knowledge that it also acted on men.[11]

The physicist, the chemist and the mathematician have their own institutions and publications through which they can express themselves; they express themselves in a language accessible only to specialists in their own field and they no longer appeal to intellectuals in general for the recognition of their work. It is in the restricted circles of specialists that the researcher finds his identity as a scientist. The divorce of philosophy and science has occurred not so much on the grounds of incompatibility of temperament as by mutual consent; the frame of reference so far common to both of them has split up, related to different occupations and environments, and has been fed with information whose criteria, channels and addresses are defined by distinctly separated intellectual communities. Whereas in the eighteenth century science was still geared to a philosophy of man, those who devoted their whole time to it already tended to treat philosophy purely as scientists. The end of science as humanism marks the end of the savant as philo-

10. Speaking of 'the almost unlimited power which scientific knowledge gains over all the thought of the Enlightenment', Ernst Cassirer noted that 'A deeper insight into the spirit of laws, of society, of politics, and even of poetry seems impossible unless it is pursued in the light of the great model of the natural sciences.' (*The Philosophy of the Enlightenment*, Beacon Press, Boston (1955), 45–6.)

11. See J. Ben-David, 'The Scientific Role: The Conditions of its Establishment in Europe', *Minerva*, 4, no. 1 (autumn 1965), 49–50. On the norms of science, see Robert K. Merton, 'Science and Democratic Social Structures', in *Social Theory and Social Structure*, Free Press, New York (1957), 550–61.

sopher and his metamorphosis into a technician of the natural sciences who no longer has anything to say, as such, about the problems of philosophy.

It was not until the middle of the nineteenth century, however, that terminology took note of this evolution—and even then with some difficulty; when Whewell proposed in 1840 that those concerned with science should be called 'scientists' instead of 'natural philosophers' or 'savants', he encountered as much opposition as approval.[12] Even more significant is the absence in continental Europe of a word corresponding to that invented by Whewell; science became institutionalised, the researcher became professionalised and the intellectual adventure of discovery became associated with the metamorphoses of large-scale industrial production, but this crossing over of scientific research on to the production side was not marked by any term designating the new function of the researcher. A. de Candolle noted in 1873 that it was 'bizarre' to be forced into circumlocution to describe 'men who search for new ideas and discoveries': 'the ordinary term of "savant" is too vast', since researchers form only a small part of the savants, that is to say, people who know (*'gens qui savent'*).[13] Savant—a man who knows; the word thus denotes a state and not a function.

By distinguishing himself from the man of learning or the teacher, the researcher allies himself with the producer, and it is precisely this new function which terminology repudiates. The reluctance so far displayed by *all* European languages except English to recognise the substantive 'scientist' is a clear sign that the divorce from culture, albeit by common consent, is still felt as something of a scandal.[14] The novelty of the

12. W. Whewell, *Philosophy of the Inductive Sciences*, T. W. Parker, London (1840), Aphorism 16, cxiii.

13. A. de Candolle, *Histoire des sciences et des savants depuis deux siècles*, Geneva–Basle (1873), 28. Candolle was so unfamiliar with the word 'scientist' that he wrote that the English language is 'even poorer' than French or German, 'since the word "learned" being thought unsuitable as a substantive, authors have sometimes used the French word *savant*, naturalised in English, "a great savant".' (*Ibid.*, 29.)

14. For example, *L'Encyclopédie française*: 'One speaks of a *scientific* treatise, in contrast with a practical work. . . . It is rarely used of people.' (Volume 14, 789). Littré clearly saw the absurdity of this usage: 'This word, which seems to have been coined in the fourteenth century, means practising science, and this is also the meaning which it has in Oresme. But in the sense in which we use it, it would be better with the termination –*al* or –*aire*: *sciental* or *scientaire*' (under '*scientifique*', volume 4 (1878 edition) 1856). The Dictionary of the French Academy (8th edition, 1935) still omits this substantive, which had to wait for Robert before obtaining final recognition.

word not only reflects a remodelling of the cultural system under which science ceases to be a benchmark, a complement or a mediation for philosophy, but, above all, it confirms the social affirmation of science as a practical activity. Scientific research is defined as a trade, science as work and the scientist as a *worker*. There is nothing surprising about the fact that, in communist language, the researcher or 'academic research personnel' is called a 'scientific worker'; science conceived as a liberal activity, distinct from manual arts and production, has said its last farewell to our modern world.

The cultural system nevertheless retains as a souvenir, like an analytical repetition of childhood, the ideology of science as a focus of values; the relation to truth prevents us from suppressing the forgotten relation to other values. There is, of course, the tradition of rationalism, for which all new knowledge is good in itself, because knowledge liberates and, by its very essence, contributes to the goals of mankind. But there is also, at another level, the revival of the exemplary value of science as consideration of truth when it is regarded not as a practical activity, a profession or a job, but as a vocation. It is the happy lot of the savant, said Fichte, 'to be destined by his particular vocation to do what a man should already be doing by virtue of his general vocation as a man';[15] by seeking the truth he achieves the whole destiny of the human race to whom he is 'teacher' and 'educator'. Fichte saw the savant as a philosopher rather than a scientist, but the fact remains that, having gained their autonomy from philosophy, the natural scientists have not renounced the moral mission he assigned to them. 'Summoned to bear witness to the truth',[16] it is through the medium of this service that the scientist serves the interests of society, not directly, since science knows no master except its own proceedings. The goals of knowledge have an exemplary value on the ethical plane; beneath the admirable features of disinterested research, science is useful in the noblest sense of the word, not as one production among others but as a creation whose object coincides with the goals of humanity.

When the affairs of the commonwealth are at stake, the scientist, as such, who contributes to the conceptual construction of truth and who submits his research to the rules of the experimental method, can do nothing because 'he has no hand in it'. This neutralism ascribed to the natural sciences reached its highest flights in positivism; it *is* positivism itself in its popular reading and heritage, the profession of faith of

15. J. G. Fichte, *La Destination du savant* (1794), Vrin, Paris (1969), 77.
16. *Ibid.*, 78.

scientists who reject all relationship between science and values, by dint of excluding from the realm of rationality everything which they do not account for. It is certainly not that of the founding father, for whom the scientific problem was not separated from the political problem; Comte could never say enough against his scientific colleagues who interested themselves only in a small sector of reality and a small fraction of science, and disengaged themselves from the rest—in his eyes, the essential. At least the clergy of the middle ages had proved themselves above religion, where the doctors of modern science proved themselves far below doctrine;[17] the fact was that 'they give preponderance to the spirit of detail over the spirit of the whole'. They looked upon their specialisation as the guarantee of the positivity of their work, but 'this strange motive, very similar to the political maxims which totally forbid free speech or free press' was above all proof of 'the philosophical impotence now proper to our learned societies'.[18] Philosophical impotence implied 'political impotence of the scientific class, and even its moral degeneration';[19] by refusing to raise themselves from their restricted field to the synthesis of the sciences, to regard their speciality not as an end in itself, but as a stage on the road towards solving the only important problem, that of social reorganisation, they reacted like

> . . . an essentially equivocal class, doomed to early elimination, as intermediaries between engineers and philosophers, without having the marked characteristics of either.[20]

Comte thought that this 'bastard character', this false position of scientists, should vanish under the twofold pressure of industrialisation and positivism triumphant. 'Most existing savants will merge with the engineers' to act on the outside world, while 'the most eminent of them will no doubt become the nucleus of a genuine philosophical class' devoting itself, by social studies, to 'the intellectual and moral regeneration of modern societies'.[21]

The process of industrialisation has proved Comte partly right, but not on the point which was nearest to his heart. If a certain positivism has triumphed, it is that displayed by most scientists who 'without any true speculative diversion' behave like pure technicians, indifferent to the ends to which their mastery of a speciality makes them contribute.

17. A. Comte, *Cours de Philosophie positive*, Schleicher Frères, Paris (1908), 57e leçon, volume 6, 256.

18. *Ibid.*, 262.   19. *Ibid.*, 273.

20. *Ibid.*, 270.   21. *Ibid.* and note 1.

Not all scientists, of course, have merged with the engineers; the divorce of modern culture, starting with a physical separation between science and philosophy, has extended into a spiritual separation between pure science and engineering sciences. When it comes to scientific research, however, there is no distinction between them; complementary and united, they form part of the same system, and however reluctant the two parties may be to set up house together, the fact remains that they are indeed a couple.[22]

And here we find the basis of the neutralism ascribed to the natural sciences: the same postulate which separates theory from practice distinguishes the scientific project and its consequences, knowledge conceived as an end and knowledge given reality as a technique. It is technology, the application of science, which assumes full and entire responsibility for the disadvantages of technical progress, not science whose real purpose is alien to that of technology. Science is pure precisely because it has no end except itself or in itself. The extension of knowledge for its own sake cannot be regarded as a means; science is neither used nor usable as such, it is *pure* as the newborn babe, without need of proof—the case is won because there is no accusation. 'God is innocent'; the problem of pure science is exactly that of theodicy. Just as God's innocence must be man's guilt, so the innocence of science must be the guilt of technology.

Thus, we read in a recent and remarkable report on the consequences of technical progress left to itself, more charged with disadvantages and dangers than with advantages:

> Scientific discovery can indeed have important consequences for the
> ethical and moral foundations of a society, but our concern here is not

22. In the United States, for example, a National Academy of Engineering Science was created in 1964 on the model of the National Academy of Sciences, to which, moreover, it is legally subordinate. The constant problem of academic institutions has been how to recognise technology, first the arts and later the applied sciences, whether by integrating them, hiving them off or sponsoring them in their own institutions. Everywhere, except to a smaller extent in the Communist countries, the link is one of subordinatiion. The problem was raised very early in France, where Colbert asked the Academy, only ten years after its foundation, to draw up the 'description of the Arts'. In 1694 Filleau des Billettes drafted a charter on behalf of the responsible commission, asking the question 'Will it be an independent Academy or the member of another?' The project was not followed up, and des Billettes joined the company five years later, in the capacity it is true of an expert in mechanics, but rather as the hostage of the arts among the geometricians. See Claire Salomon-Bayet, 'Un préambule théorique à une Académie des Arts', *Revue d'Histoire des Sciences*, P.U.F., Paris (July–October 1970), 23, no. 3.

with the effects of science—what man knows or hypothesises about his world—but with the effects of technology—what man can do and chooses to do with what he knows.[23]

Case dismissed: the case is closed before it has even been pleaded. It is evident that if the *effects* of science are knowledge and hypothesis, there is no point in summoning it to appear in the dock; just as God has no hand in the stumbling-block of evil, so science has no hand in the stumbling-block of its applications.

> If it is true, in Kant's words, that reason has never ceased to bring charges against the supreme wisdom, pointing to everything in the world which runs counter to good, it is none the less true that it is also reason which, at the risk of being divided against itself, has been constantly employed in clearing God from these same charges.[24]

In the same way, the scientific community must be divided against itself, the grain of the researchers must be separated from the chaff of the engineers, if the cause of science is to be won without a hearing. Death of the savant, apotheosis of the scientist: turning his back on all philosophy, he nevertheless retains its garb, since, if he becomes a technician, it is, in his belief, a technique which has no effect other than the conceptual. The natural sciences have no sooner abandoned ideology in the name of their scienticity than they rediscover ideology in their scienticity itself.

The emergence of a political species

The end of 'laissez-faire' in the relations between knowledge and power nevertheless means the emergence of scientists as an ambiguous species of the genus political animal: an ambiguous species, first, because it denies belonging to that genus. Scientists, says C. P. Snow, have the future in their bones;[25] for them there are no locked doors, all doors open to a well-made key. Or again, as Leo Szilard says, since their quest is for clarification and truth, the insincerity and dubious motives of the traditional actors on the political scene are not their lot.[26] In sum, they form a particular species of the human genus, and their image of themselves and the sensitivity and intellectual attitudes on which it is based

23. *Technology: Processes of Assessment and Choice*, Report by the Committee on Science and Public Affairs of the National Academy of Sciences, Washington (1969), 9.

24. Jacques Brunschwig, introduction to *Essais de Théodicée* by Leibniz, Garnier-Flammarion, Paris (1969), 8.

25. C. P. Snow, *The Two Cultures*, Cambridge University Press (1959), 10.

26. Leo Szilard, *The Voice of the Dolphins*, Simon and Schuster, New York (1961), 26.

reflect a relation with the affairs of this world which has no equivalent elsewhere.

> The scientist [says Warner B. Schilling] is likely to find little in common with the diplomat (who is inclined to believe that most of his problems have no solution and who is in any event too busy with the crises of the day to plan for tomorrow), or with the politician (whose approach to problems is so spasmodic as to seem neither analytical nor rational, and whose policy positions are anyway soon blurred by his efforts to accommodate to the positions of others), or with the professional student of international politics (who, when the opportunity permits, lectures the scientist on the elegant complexity of the political process, but who never seems, to the scientist at least, to have any really good ideas about what to do).[27]

Why should the methods and attitudes which have proved themselves in the scientific field not apply with equal success in the political field? The scientists' image of their political aptitudes relates back to the eighteenth century model which postulated a close correspondence between natural history and human history in so far as both were ruled by the same rational approach; if this approach is transferred from one kind of history to the other, there is every hope of overcoming the contradictions of the *politeia*. It is the specific qualities of scientists which justify the claim, in the Vienna Declaration of the Pugwash Movement, to special competence in the affairs of this world.[28] And even more, the whole content of the scientific enterprise may inspire the solutions of conflicts between individuals and nations. Thus, Max Born and Oppenheimer refer to the theory of complementarity as an experience of compromise which could be transposed into the political field; the ability of physicists to adopt new forms of thought to save their discipline from apparently insuperable crises should be a model to follow in human relations.[29]

27. Warner B. Schilling, 'Scientists, Foreign Policy, Politics', in *Scientists and National Policy-Making* (ed. Gilpin and Wright), Columbia University Press, New York (1964), 154–5.

28. 'Scientists are, because of their special knowledge, well equipped for early awareness of the danger and promise arising from scientific discoveries. Hence they have a special competence and a special responsibility in relation to the most pressing problems of our times.' (Declaration of Vienna, 3rd Pugwash Conference (14–20 September 1958), in J. Rotblat, *Pugwash: A History of the Conferences in Science and World Affairs*, Czechoslovak Academy of Sciences, Prague (1967), 95.)

29. 'I should like to try to apply to politics the proven method of physics which consists in radically changing your point of view; this would reveal to us compatible and complementary situations in those oppositions which are deemed to be irreconcilable.' (Max Born, *La responsabilité du savant*, Payot, Paris (1967), 143.) Oppenheimer has on several occasions referred to Niels

Among all forms of expertise, that of the scientist seems best equipped to neutralise the element of passion in social debates and to propose purely technical solutions; the scientific method, if not the scientific spirit, should be able to reduce the equivocal terms of action involved in conflicts of values to the objective findings of unprejudiced judgments. In theory, everything inclines the scientist to eliminate from his judgment the non-technical preconceptions which mark the limits of human decisions: the temperament which leads him to research work, and induces him not to despair of the problems he handles and to take it for granted that they have a solution; the method to which he constrains his work, involving rigour, attention to facts, respect for criticism and proof; and, finally, the operational character of science itself, whose objectivity is guaranteed by the function of measurement in the definition and construction of the objects to be measured—independent both of the will of the experimenter and of the frame of reference adopted, the phenomenon achieved in a rational programme of experiments is protected, in Bachelard's words, against any irrational disturbance.[30]

This operational character of modern science explains why it tends to solve sharply delimited and finite problems; even the most general theory no longer aims at being universal.

> One of the paradoxes of modern science [says Harvey Brooks] has been that the greater its success in a pragmatic sense, the more modest its aims have tended to become in an intellectual sense. The goals and claims of modern quantum theory are far more modest than those of Laplace, who believed that he could predict the entire course of the universe in principle, given its initial conditions.[31]

It is doubtful whether, as was said somewhat hastily in the 1950s, the reign of technical competence has tolled the knell of ideologies, but it is true that the idea of a universal system of interpretation and action has not withstood the assault of the results unceasingly accumulated by the pragmatic approach of science. This approach is all the less apt to

---

Bohr's theory as a model of apprenticeship to toleration; for example: 'This discovery has not, so far as I know, penetrated into general cultural life. I regret this; it is a good example of something which could be applied in depth, if only we could manage to understand it.' ('Science and Culture', in *Les Etudes philosophiques*, P.U.F., Paris, no. 4 (October–December 1964), 531.)

30. See G. Bachelard, *La Philosophie du non: Essai d'une philosophie du nouvel esprit scientifique*, P.U.F., Paris (1940), 4–11, and Jean Ullmo, *La pensée scientifique moderne*, Flammarion, Paris (1958), 17–40.

31. H. Brooks, 'Scientific Concepts and Cultural Change,' *Daedalus*, Cambridge, Mass (winter 1965), 66–7.

cement the construction of an ideological synthesis, since its pattern is more and more that of fragmented and insulated work.

> One sees a close analogy between the preoccupation of science with manageable problems and the decline of ideology and growth of professional expertise in politics and business.[32]

And yet ideology is not dead, far from it; the attempt to free social issues from passion by analysing them scientifically might well be regarded as a substitute for ideology. Stripping these issues down to their technical aspects does not get rid of the conflicts of ideas and values from which they spring. They cannot be fitted into an overall synthesis without some dogmatism, but it is no less dogmatic to expect the proceedings of expertise to finalise the endless tale of the philosophies of history. This is manifest today in the United States, where the apparent triumph of technical solutions and the technocratic optimism of the 'new society' have culminated in an ideological clash which, in the context of the Vietnam war, sustains the challenge launched against the 'new mandarins' by radical thought.[33] Already at the first trumpet blasts which were to sound the knell of the 'ideological age', Raymond Aron humorously noted that this controversy, however different its sense in the United States, could boast of a long history in Europe:

> In turning back from ideology, the American anti-ideologists did not turn very far; some of them merely turned back from Europe.[34]

The guarantee of objectivity from which science benefits is all the less applicable to politics because the latter, whatever the success of scientific procedures, is never through with ideologies.

Scientists are an ambiguous species for the further reason that it is hard to assess the nature and weight of their influence. Some people regard their potential role as considerable, even if in practice the political leaders take no notice of their advice or misinterpret it through inattention or incomprehension; if they were to follow that advice, political affairs would take a quite different turn and all the problems of life in common, of relations between individuals and between nations, would be solved as scientific problems by the sound rules of non-contradiction, experiment and verification.[35] Others regard this tendency to pontificate in matters beyond their technical competence as

32. *Ibid.*

33. See, in particular, Noam Chomsky, *American Power and the New Mandarins*, Pantheon Books (1969) and Penguin (1969).

34. R. Aron, 'Fin des idéologies, Renaissance des idées', in *Trois Essais sur l'age industriel*, Plon, Paris (1966), 198.

35. C. P. Snow: 'Scientists have it within them to know what a future-directed society feels like, for science itself, in its human aspect, is just that.'

a flaw rather than a characteristic, sometimes a virtue of innocence like that of the absent-minded professor in the comic strip, sometimes a sin of arrogance on the model of Doctor Faustus; scientists at best are inoffensive by reason of their naïvety and lack of realism, at worst they form the élite which, in President Eisenhower's words, threatens to lead democratic institutions captive.

They certainly form an élite, but not in the sense of the Russian 'intelligentsia' of the nineteenth century, a thin layer of enlightenment, whose interest in public affairs was all the greater because the majority of the population was in no state to take part in them. Not only do the scientists form a community of impressive and ever-growing size, but, however reluctant they may be to take part in public affairs, their activities are not irrelevant to the conditions of life of the economic and political systems in which they are conducted. If, unlike the Russian intelligentsia, they took no interest in public affairs, they would still influence their course. The paradox of this new intelligentsia is precisely that it is fully entitled, by virtue of the values it proclaims, to turn its back on the political scene while, in the name of its own interests and its influence over the interests of power, it plays a part which is no longer merely that of a walker-on.

That is why nothing could be more inaccurate than to regard it, with Robert Wood, as

> . . . an apolitical élite, triumphing in the political arena to the extent to which it disavows political objectives and refuses to behave according to conventional political practice.[36]

Wood is inspired by Harold Lasswell's theory of directing élites, and particularly by the model of 'skill commonwealths' constructed by Lasswell to account for 'influence groups' in the light of criteria based not on social status, personality, or geographic, racial or religious origin, but on the exercise of a common range of skills which can be taught and learned.[37] No doubt this model is convenient for distinguishing scien-

---

(*Science and Government*, Mentor, New York (1962), 73.) Or in the east, recognising 'how far politics, economics, the arts, education and military questions still are from being handled with the aid of the scientific method', Andrei Sakharov introduces his essay with the statement that the problems of the modern world 'demand the use of this method'. (*La Liberté intellectuelle et la Coexistence*, Gallimard, Paris (1969), 29–30.)

36. R. C. Wood, 'The Rise of an Apolitical Elite' in *Scientists and National Policy-Making*, edited by Gilpin and Wright, Columbia University Press, New York (1964), 48.

37. *Ibid.* 51. See Harold D. Lasswell, *The Analysis of Political Behaviour; an Empirical Approach*, Oxford University Press, New York (1948).

tists as a group from other 'power élites'—especially lawyers. There
are many more lawyers than scientists acting as government advisers
and administrators; furthermore, the skills proper to research and the
relation to facts inherent in the scientific approach do not seem calculated to give scientists any advantage over lawyers in political debate:

> Men skilled in the manipulation of symbols of things confront men skilled
> in the manipulation of symbols of interpersonal relations.[38]

Logic against dialectics; the very qualifications which some scientists
rely upon as making them more competent than other social groups to
deal with political problems should rather disqualify them from the
political arena—as though the art of preaching and their biblical knowledge of history should have stopped the priests of the middle ages from
rivalling the warriors.

The disadvantage of this approach is that there are hardly any data
to uphold it or refute it. Empirical sociology has so far taken little
interest in studying the relations of scientists with power or even their
political behaviour.[39] But the question goes deeper than this; the definition of élites by their 'skills' disregards the functions they discharge in
this relation. The lawyer may well be more skilful than the scientist in
the practice of interpersonal relations, hold more administrative jobs,
or have more direct access to the top executives, and still carry less
weight than the scientist on questions where the scientist has greater
skill, and even on questions where he has less.

> Institutional position is important because of the power it confers [says
> Pierre Hassner, very properly], and is largely explained, together with the
> nature and limits of that power, by the specific or overall function of the
> institutions in question in the state or in society.[40]

38. Wood, *op. cit.*, 51.

39. Empirical inquiries into the political opinions and activities of scientists
are not only rare, they are above all limited in scope, by reason both of the small
number of interviews and of the restricted themes they adopt. See Anne Roe,
The Making of a Scientist, Dodd, Mead, New York (1952), and Warren O.
Hagstrom, *The Scientific Community*, Basic Books, New York (1965). The best
inquiries are those which have been the subject of two still unpublished theses
presented to the Department of Political Science of the Massachusetts Institute
of Technology: David Nichols, *The Political Attitudes of a Scientific Elite*
(1968), on American scientists, and Albert H. Teich, *International Politics and
International Science* (1969), on European scientists and intergovernmental organisations. The reference book in this field does not belong to empirical sociology, but rather to political science and even political philosophy; it is Robert
Gilpin's *American Scientists and Nuclear Weapons Policy*, Princeton University Press, New Jersey (1962), whose analyses have largely inspired our discussion.

40. P. Hassner, 'A la recherche de la classe dirigeante: le débat dans
l'histoire des doctrines', *Revue française de Science politique*, P.U.F., Paris, 15,
no. 1 (February 1965), 46.

From the outset, emphasises Pierre Hassner, this debate is itself ideo-logical:

> Judging it is a matter of empirical inquiry, but the orientation and signific-ance of the inquiry, the stress on power or function, will obviously depend on the conception which is adopted of society as a whole and its relations with the different groups as much as on the analysis of each group.[41]

Even if their relations with power are judged solely by the criterion of skills, it is far from evident that scientists form an 'apolitical élite'; like sworn assessors advising a tribunal, their opinion carries weight solely because of its own objectivity and not because of some function it may fulfil in the institution to which they are attached. The success of expertise as an objective way of proceeding, formulated 'above the fray', naturally depends on how far the expert is able to free himself from pressures or 'conventional political practice'. But the whole ques-tion is how far the function of an expert in political matters can be assimilated to that of an expert assessor to a tribunal, especially in the case of a scientist, who, even in spite of himself, is an interested party in the operations about which he tenders his advice.

Expertise and responsibility

When a scientist regards himself or holds himself out as being able to find solutions which are bound to carry conviction with his hearers purely by virtue of being founded on fact and on the rigour of scien-tific reasoning, he overlooks what Robert Gilpin recalls, that

> even in scientific research where rigorous conditions exist to encourage objectivity, it is impossible for the scientist to free himself from his values and his implicit assumptions of a non-technical or non-scientific nature.[42]

Still more so if he is dealing with scientific problems in the political field or with political problems in the guise of scientific expertise; no more does he escape from the ambiguities of conflict of interest when, speaking in the name of science, he merely respresents in fact only a narrow sector of research activities than he can shake off his personal values and assumptions by placing his opinion under the sign of scien-tific reasoning. The passion for objectivity itself is merely one passion the more:

41. *Ibid.*
42. R. Gilpin, *American Scientists and Nuclear Weapons Policy*, Princeton University Press, New Jersey (1962), 22.

This emotional attachment to one's own point of view is particularly insidious in science because it is so easy for the proponent of a project to
clothe his convictions in technical language.[43]

If the scientific process had the same force of persuasion in its relations
with power that it has in its relations with truth it should not arouse
(or at any rate not to the same extent) the controversies to which it is
nevertheless constantly exposed.

But to overcome the antinomies of the 'ethic of responsibility' it
is not enough to profess to extend outside the sanctuary of truth the
methods and attitudes on which that truth is founded. Truth cannot be
transplanted on to political ground without losing everything which, in
the eyes of the scientist himself, ideally defines it as the object of universal agreement. This was quite clear from the differences about nuclear
fall-out, such as those which set Ernest O. Lawrence and Edward Teller
in opposition to Linus Pauling, in spite of the fact that they had all been
briefed from the same dossier: within the limits of the same known
facts, the threat of fall-out amounted for one side to a few dozen people
with innocuous effects, for the other to millions with catastrophic
effects.[44] The clock on the front of the *Bulletin of the Atomic Scientists*, the symbol of the last few minutes of mankind's eleventh hour,
in the absence of international agreement on atomic weapons, does
not—happily—count the hours in mathematicians' time.

These partialities of the scientists are so manifest in their approach
to foreign policy that they seem to be specific to them, as though the
application of the reputedly objective practices of science resulted in
consistent behaviour. Thus, Warner R. Schilling has identified some of
their characteristic reactions, especially those of the physicists faced with
problems of national security.[45] First there is 'naïve utopianism or naïve
belligerency' which looks for a simplistic and mechanistic solution to
international relations; but there is also the desire to approach a problem with full knowledge and to subordinate its solution to an exhaustive
study of all its aspects. Hence the difficulties which arose between the
military and the scientists at the time of the Manhattan Project, the
military fearing that if scientists were allowed to range over the whole

43. James Conant, *Modern Science and Modern Man*, Doubleday, New
York (1952), 114, cited by Gilpin, *ibid.*

44. See *Le Monde*, 12–13 May 1957, for Lawrence, and 4 June 1957, 6,
for Pauling.

45. W. R. Schilling, 'Scientists, Foreign Policy, Politics (The Problem of
Bias)' in *Scientists and National Policy-Making*, 152–63.

problem, they would never find a solution in time.[46] Similarly, the idea of a radical leap rather than a series of modest steps forward in the solution of technical problems is extended into the idea of a radical solution rather than a progressive improvement in international relations—as though the epistemological myth of 'the crucial experiment' generated, when science is applied to diplomacy, the myth of the final and decisive solution.

Even more significant is the situation in which the scientist, blinded by the beauty of the technical solution he has in mind, succeeds in eliminating the political context of the decisions which determine the urgency or the nature of that solution. This 'complex' of 'technology for its own sweet sake', or, in Oppenheimer's words, the 'technically sweet' solution, which makes it seem imperative to complete a project once it has been deemed feasible, no matter what the consequences, appears to Schilling, not without reason,

> . . . to have its roots in two more of science's central credos: the belief in the value of pursuing knowledge for its own sake, and the belief that the best motivation for the direction of research is the strength and character of individual curiosities.[47]

The complex of 'technology for its own sweet sake' leads to the suppression of everything which links the problem raised and its solution to the political environment and gives them their true significance in relation to that environment. The assertion of neutrality in this context assumes its most aggressively deceptive form, as though the aesthetic beauty of the technical solutions had nothing to do with the use of the tools which are forged with their aid.

The expert assessor to a tribunal is not expected to intervene in the field in which he is arbitrator, whereas the scientist manipulates his own field even when the pleasure of finding the technical solution renders him blind or insensitive to the consequences; this aesthetically induced blindness does not make his intervention any less real, it does not change the instrumentality of knowledge. Science for science's sake is not art for art's sake, unless, of course, one rules out all consideration of the end which it serves in satisfying its own ends.

46. Justifying his efforts to 'compartmentalise' research at Los Alamos, General Groves was to declare that he was afraid not only of security leaks but of 'perfectionism': 'If I brought them [the scientists] into the whole project they would never do their own job. There was just too much scientific interest, and they would just be frittering from one thing to another.' (*In the Matter of J. Robert Oppenheimer: Transcript of Hearing before Personnel Security Board,* USGPO, Washington (1957), 164.)

47. W. R. Schilling, *op. cit.,* 160.

It is my judgment in these things that when you see something that is technically sweet you go ahead and do it and you argue about what to do about it only after you have had your technical success.[48]

In these famous words of Oppenheimer we find all the equivocation and all the insincerity of scientists in their relations with power. By asserting that the object of research is good on the sole ground that it is feasible, and all the more feasible because it is 'technically sweet', they fancy they can divorce their expertise from its implications, declining all responsibility in the name of the requirements of research—aesthetic requirements in this specific instance, whose apparently non-instrumental character is nevertheless what generates the instrumentality of the product.

Scientists are also an ambiguous species because it is difficult to determine who represents them and who speaks in their name. Empirical sociology might perhaps show, in carefully processed statistical terms, that it is hazardous to argue about their political attitude, except case by case or by types of case which cannot be expected to be generally applicable. Indeed, the research system, ranging from university scientists to engineers in public and private laboratories, is not manned by a homogeneous population, and free researchers, however much they may be localised in the universities, are no more homogeneous. The idea of the 'scientific community' calls up the picture of a professional group united by the similarity of its intellectual interests and the norms by which it is guided. But while it is true that scientists share a common intellectual area, this area is never self-consistent, and the links which hold it together operate only on the ideal plane of its own specific values. But more than that: under Nazism or Stalinism, as under the cultural revolution (one has only to look at the image of western science presented in Chinese scientific journals), these values carry very little weight compared with political fanaticism. Whatever their formal or ideal links, the members of the 'scientific community' represent activities, professions, status, institutions and nationalities which are too different to enable them to speak or to understand each other with a single voice. The republic of savants is neither one nor indivisible and

48. This comment, originally made about the first atom bomb, was repeated in connection with the H-bomb, where Oppenheimer disagreed with Teller because he did not think it was feasible: 'Let us not talk about use. But my feeling about development became quite different when the practicabilities became clear. When I saw how to do it, it was clear to me that one had at least to make the thing. . . . The programme in 1951 was *technically so sweet* that you could not argue about that.' (*In the Matter of J. Robert Oppenheimer, Transcript of Hearing*, 81, 251. My italics.)

no more constitutes a united family than any other professional group. As we shall see below, even the fact that the area of scientific activity has no frontiers does not give the 'international scientific community' any more coherence than any other.

The scientific community is based on the idea of Reason in Kant's sense; it is a community whose activity is universal as of right and which, if it exists as an idea only, is nevertheless regulated by usage. In reality, there are a number of scientific communities whose institutions and representatives are organised in varying degrees in their relations with power. This distinction is justified by the fact that the state takes a greater interest in certain disciplines or fields, and their representatives for this very reason form more or less influential pressure groups; the physicists are better organised than the chemists, the molecular biologists than the mathematicians, the natural sciences in general better organised than the social sciences. But it is also justified by national structures and traditions which lead to particular collective reactions to the same problem; American scientists are better organised than their European colleagues, if only because the adventure of the first atom bombs brought home to them earlier and more acutely the consequences of the relation between science and politics. Thus, the Federation of American Scientists, both meeting place and platform, played a decisive role in the crusade of the Chicago and Los Alamos atomic scientists to sensitise public opinion (and especially Congress) not only to the issues at stake in connection with the new weapons, but also to the importance of scientific research.[49] The experience of the second world war and of competition with the Russians has resulted in the United States in a dialogue between scientists and politicians which has no equivalent in any western European country.

The national scientific communities certainly do not carry the same weight in every country. Taking the example of France, which moreover is repeated in most European countries, researchers, being officials, are represented by trade unions in contact with the administration rather than by lobbies in contact with the executive and the legislature. The very difference of institutions and political customs entails a difference in the nature and action of scientific pressure groups. In spite of these differences, however, the creation of national bodies responsible for formulating science policies has everywhere made scientists aware

49. See Alice Kimball, *A Peril and A Hope: The Scientists' Movement in America, 1945–47*, University of Chicago Press (1965), which gives the best report of the birth and role of the atomist lobby.

of the negotiating power they hold, a power which the traditional scientific institutions had allowed to lie fallow, even if from their origin they were assigned, like the Academies, the mission of exercising it. Whatever the discipline or the country, the new relation instituted between science and power has the same foundation and the same consequences; science finds an inevitable sleeping partner in the state, and the state finds the active partners it needs in the scientists. At the same stroke, the recognition of science as an object and issue of politics leads the scientists to recognise themselves as subjects of politics. Scientists experience the scientific community as an ideal community; politicians regard it as, or would like it to be, an organised group. It is through the eyes of power that it comes to look on itself as a representative institution.

If the function of scientists, as much as their institutional position, prevents us from speaking of them as a community detached from political contingencies, pressures and commitments, it is because the two cannot be separated; the function governs the institutional position and the institutional position governs the discharge of the function. Wood's interpretation puts forward the power of technocracy as it is commonly denounced by raising the bogey of government by technicians, but where has there ever been a government in which the experts 'triumphed' by virtue of disavowing 'conventional political practice'? There seems to be greater relevance in the formula of Don K. Price, who regards scientists as

> . . . an *establishment*, in the old and proper sense of that word; a set of institutions supported by tax funds, but largely on faith. . . .[50]

Price immediately adds, however, that these institutions are 'without direct responsibility to political control', which is less evident than his book tends to affirm. No doubt scientists do not constitute an 'estate' in the sense of the Third Estate, claiming to be recognised in the division of power, or an organised group with the 'technostructure' seeking to engross it. Following the religious pattern, Price envisages a sort of division of labour: just as the sacred is distinguished from the profane, so science, the pursuit of truth, is distinguished from politics, the pursuit of power. And the guarantee of the system of 'checks and balances' applied to their relations is that there should be no interference: the more the purpose of an institution is truth, the more it

---

50. Don K. Price, *The Scientific Estate*, The Belknap Press of Harvard University Press, Cambridge, Mass. (1965), 12.

deserves to be free and left to govern itself; the more its purpose is power, the more it should be subject to the test of political responsibility, that is to say, in the last instance, to the decision of the electors.[51]

This projection of the American constitutional system into the relations between knowledge and power is perhaps valid as a principle to guarantee academic freedom, but it presupposes both a science which is not power and a power which is not interested in science. In practice, far from bowing to the system of 'checks and balances', the principle of the relation which now exists between science and power is that of reciprocal intrusion. The sacred meddles with the profane, science with politics, and it is exactly this marriage of the dissimilar which is one of the sources of unrest in university research, so dependent on the interested largesse of the state that the profane seems to have wiped out the very memory of the sacred. Just as the existence and fortune of the establishment depend on the good will of power, so the good will of power depends on the services rendered to it by the establishment; the sacred disclaims the profane, but in reality it is already secularised.

It is obviously in the sphere of strategy that the apoliticism of the scientist appears the least plausible. It is no accident that people talk about 'weapons systems', that is to say, the integration of hardware and software, firepower and brainpower, studied by soldiers, mathematicians, physicists and biologists in association with economists, psychologists, sociologists and philosophers (the 'think tanks' such as RAND or the System Development Corporation were the birthplace of the theory of 'integrated systems'). Science appears here in the pure state as a technique, an instrument which cannot be dissociated from the social system which gives it meaning, defined both by the power it confers and the function it fulfils.

The decision at the start of World War II to develop a fission bomb [recalls Albert Wohlstetter] or the decision to use it against Japan, or the decision to develop an H-bomb, or to bomb German cities during World War II, called for much more than natural science and engineering. Such decisions have narrowly technological components, but they involve just as essentially a great many other elements; military operations and counter-operations by an enemy, the economics of industrial production, the social and political effects of bombing on populations, and many others. Some of these other factors are qualitative. Many are quantitative, and in this very broad sense, 'technical'. (They involve numbers and may be related in a numerical model.) However, even these latter do not fit into

51. *Ibid.*, 137.

any of the narrowly technical traditional disciplines of natural science or engineering.[52]

The scientist as counsellor of power comes in as the spokesman for the community of science, without preconceptions, but his expert advice may not be confined to the technical aspects of the questions he is discussing. He may believe that, having gone part of the way only, his function—or his responsibility—ends there, but he can nevertheless not dissociate his advice from its political implications. In strategic matters, expertise not only consists in defining the technical problem which is the subject of decision, but it must also bring into the open the political implications underlying the technical problems. The negotiations on the nuclear test ban provide an illuminating example.

> In the early days . . . both scientists and diplomats fell into the trap of believing that the basic issues were primarily technical ones which could be resolved by discussions among experts, if not at the time, then later on as new scientific knowledge became available. Subsequently, it became increasingly clear that the really difficult issues were related to the degree of assurance which the United States felt it must have against the conduct of clandestine underground tests, and to the Soviet judgments as to what would be the acceptable degree of penetration of their military security.[53]

There could be no better description of the candour of scientists faced with a problem whose solution, however technically evident and convincing, depends on implications and judgments in which technical consideration were either irrelevant or into which they introduced new data which changed their significance:

> The importance of detecting clandestine underground tests has been differently estimated in the United States, depending on judgment as to the military decisiveness of tactical nuclear weapons.[54]

It may be noted in passing that, in this particular instance, the naïvety of the scientists was equalled by that of the diplomats, less accustomed as they were, on the pattern of 'communities of skills', to the idea that

52. A. Wohlstetter, 'Strategy and the Natural Scientists', in *Scientists and National Policy Making*, 178. See also McGeorge Bundy: 'You don't solve the problem of nuclear weapons and their relation to the world by saying "Here is a nuclear core—that's scientific; here is a nuclear weapon—that's military; here is a treaty—that's political." These things all have to live with each other. There are elements that are indeed military or technological or diplomatic, but the process of effective judgment and action comes at a point where you cannot separate them out. It follows, I think, that it is also nonsense to talk about the political neutrality of scientists'. ('The Scientist and National Policy', in *Knowledge and Power* (ed. Sanford A. Lakoff), The Free Press, New York (1966), 429.)
53. H. Brooks, 'The Scientific Adviser', *ibid.*, 85.   54. *Ibid.*

every problem must have a solution. It is true that this was an extreme situation in which the strategic issues at stake might well blind the clearest of minds, particularly since it was entirely novel in the nature of the technical factors involved as well as in their political implications; how can you rely on technical arguments when you cannot or will not rely on your opposite number? Price remarks that while Einstein's saying that 'God is subtle, but He is not malicious' is an encouragement to those scientists who seek to find order in the complex universe, in an international negotiation on, for example, arms control, no matter how high a scientific content the subject matter may have, the contribution that science can make is a limited one, because the other side may be malicious.[55]

Now, even when the issues are less dramatic, the relation to politics is such that recommendations based on technical considerations alone are just as much based on premises and lead to consequences which go beyond the apparently neutral field of expertise. It is true that this encroachment of politics into the expertise of scientists can be tempered according to the nature of the issue and the parties involved in the decision; if the issue is one of health and not of weapons, of universities or industrial innovation and not of nuclear or space research, and if it is debated at the level of a technical ministry or a public laboratory and not at the summit of the executive, it is self-evident that non-technical considerations may play a smaller part. But the fact that a technical question does not directly affect the state in its capacity as legitimate monopolist of violence does not thereby dispel the political environment in which such a question affects power or is affected by it. This is firstly because a question ceases to be purely technical as soon as it calls for a decision; neither the fact that the advice is tendered by a scientist nor the fact that it is couched in scientific terms justifies it in disregarding the underlying presuppositions, any more than advice from any other source. (It would be no less naïve on the part of the politician to rely on expert advice solely because it is based on a scientific analysis of the question in issue.) Secondly, scientific questions which become questions of power immediately come under the common rule of the political scene; the objective and rational space of scientific discourse is inevitably apprehended as the object of political discourse by all the actors involved, including the scientists.

55. Don. K. Price, *The Scientific Estate*, The Belknap Press of Harvard University Press, Cambridge, Mass. (1965), 152.

S.P.—7

Technonature and the end of innocence

All scientific activity is in fact a question to power; *the rapprochement between theory and practice does away with the cleavage between ideology and the scienticity of the sciences which was the basis of the positivist claim for the neutrality of scientific discourse.* So long as theoretical activity could be separated from its applications this discourse was no more than intellectual, indifferent to events which were not part of its own history: science talking about science, and not in relation to world history. But the theoretical function of science can no longer be dissociated from its practical function: in fulfilling itself as technology, the discourse of science has ceased to be pure, except in so far as it is conscious of the way it has been determined by the actual course of history, instead of remaining unconscious of it, forgetting it or suppressing it and in so far as it recognises that this determination is the reason for the role it plays in world affairs. In this way it can plead that it is not the cause of the consequences it involves: it does not look upon itself, like technology, as an instrumental means to ends which are alien to it; it is not the cause but only the pretext and the occasion of the development which it generates; it is not 'answerable' for them.

Precisely because of its scientific character, science cannot be regarded in the equivocal role of an instrument, as technology can. Its truth lies not in its latent function as an instrument of power, but in the deep roots of an effort of theorising in which conception is divorced from practical application and has no mastery over it by dint of pursuing its own history in isolation from techniques. The whole concept of 'pure' science aligns the theoretical attitude and the moral attitude as a field apart in the research system, a radical frontier between the 'ideological' and the 'scientific' which rules out any intrusion of political restraints into the protected citadel of knowledge. And yet, even if 'pure' science aims at no special practical result, it cannot ignore the fact that it leads to information and products on which power may lay its hands as tools and which will affect the nature and process of its decisions. And, in practice, it ignores this fact all the less since its prospective applications, however remote, are founded in its eyes on the legitimacy of the support which the state owes it. The universe of 'pure' science is not immune from the social convention which demands that every human activity must have an instrumental purpose, and all the less so since, as the source of man's mastery over nature, both its advancement and its results depend on the attention paid to it by power.

The influence of scientists within the 'technostructure' arises out of the function they fulfil and the institutional position they occupy as the possessors of highly specialised knowledge, access to which, unlike that of other members of the 'technostructure', is barred to the profane. If power in the modern industrial system, according to Galbraith, has shifted from the individual to the group, the scientists form a group apart in so far as they hold the monopoly of a form of specialised knowledge which cannot be transmitted to those who are not trained for it. It is this monopoly of specialised knowledge, much more than their direct participation in decisions, which justifies us in likening scientists to the political and religious pattern of the priesthood, a pattern which is so manifest that it is found even in a highly official report to the United States Congress:

> In a sense, the Pharaohs had the same problem as the Congress of today; how to achieve credible communication with a knowledgeable cult of specialists whose social contribution was undeniable, but whose ways were obscure and whose findings were not subject to proof test outside of the priestly cartel. Under these conditions the scientist is—or may be thought to be—subject to the temptation of advising not only on *what* but as to *whether*.[56]

Naturally, the more strictly personal to himself the scientist's knowledge, the greater his influence; it is from this angle that the brain drain has become as important strategically as certain raw materials or sources of power, but it is from this angle also that the 'scientific estate' holds a privileged position in the 'technostructure'.

> In Los Alamos, during the development of the atomic bomb [writes Galbraith], Enrico Fermi rode a bicycle up the hill to work; Major-General Leslie R. Groves presided in grandeur over the entire Manhattan Project. Fermi had the final word on numerous questions of feasibility and design.[57]

The concept of 'technostructure' so happily forged by Galbraith designates the community of all those who, as holders of technical information and knowledge, take part in the decisions of the modern industrial system founded on the organised use of capital and technology. In this community 'the educational and scientific estate' is distinguished from the others essentially by its motives:[58] its members

56. *Technical Information for Congress*, Commission for Science and Astronautics, Chamber of Representatives, 91st Congress, First Session, Washington (25 April 1969), 12.

57. J. K. Galbraith, *The New Industrial State*, Houghton Mifflin, Boston (1967), 66.

58. Galbraith notes that the idea of 'estate' is borrowed from Don K. Price's *The Scientific Estate*.

are not looking for power for its own sake or for wealth for its own sake; as agents of social innovation within the universities, as critics of power, they have interests which have nothing to do with those of the organisations on which they nevertheless depend.

> Educators, in pursuit of political interest, differ from others only in the impression of exceptional purity of motive which they are able to convey.[59]

Less zeal for the exercise of power, less greed for money, but at the same time, less conformism in social behaviour: the distinction is a fine one—not that it is not in a certain sense essential if you think, as we do, that in a society dedicated to production and profit everything which seems to make intellectual activity non-utilitarian must be religiously safeguarded. But it is based on reality only up to a certain point—or, if you like, it seems almost too late to invoke the distinction; it is immaterial that the quest for power (or wealth) is not the aim of the seekers after truth, the fact remains that in looking for the truth they encounter power (or the power of wealth).

Galbraith himself is so conscious of this that he has to attribute the subtle alienation of which educators and scientists are victim to the genius of the industrial system—an evil genius which spares absolutely nothing to encourage beliefs which justify the acceptance of its goals. If the influence of educators and scientists meets the needs of the industrial system, 'this does not necessarily create a primary obligation to the needs of the industrial system'.[60] But the needs of the industrial system do not drop as the gentle rain from heaven; they are, on the contrary, closely bound up with the information and results which science produces in the course of satisfying its own needs, thanks to the industrial system. In the 'technostructure' the representatives of the natural sciences—and with them those of the social sciences whose knowledge guarantees or promises the same power of action—owe their influence to the nature of their disciplines. Organisation and planning did not invent science, science invented organisation and planning, and the race for technological innovation can hardly be conceived as alien to the activities of the scientists; the evil genius of the industrial system was cradled in the whole essence of modern science.

We propose to coin the word *technonature* to denote the manipulation of natural forces in the light of political decisions which is now characteristic of the situation of scientists in society. Technonature is the area within which the interests and attitudes of scientists are inexor-

59. J. K. Galbraith, *ibid.*, 349.   60. *Ibid.*, 381.

ably bound up with power by responsibility for its needs and subservience to its goals. The word denotes not so much a group, a class or an élite—which, indeed, is not monolithic either in its interests or in its attitudes to power—as the ground on which the 'ideological' and the 'scientific' are allied as instruments at the service of power. Technonature is not a neutral arena: it is by no means indifferent to the ends it serves, unless it is held that the practice of science can dissociate its own ends from the ends which are not only achieved but actually conceived with its help. Scientists are, in fact, the only technicians who can act on nature itself, can offer to change its state and conditions, and can determine information and products whose novelty in turn transforms the terms of the scientific process. *It is science which has the initiative in the continuing creation of technonature as the source of new political problems.*

The atom bomb is the outstanding example of the power of technonature to generate a new situation from a conjunction between the pure concepts of science and the interests of the state. But it would be a mistake to regard this example as unique, however exceptional it appeared at the time because of the vastness of its consequences. From the university laboratories to those of public or private institutes, all fields of scientific research—from the atom to space, from the sciences of materials to the life sciences, from the most abstract calculation to the behavioural sciences—now fall within the area of technonature; the conjunction of scientific activity and political activity no doubt has preferential fields, which owe their favour both to the pressures of the current political situation and to scientific vogue, but it does not operate exclusively within those fields. It is the whole of science, conceived as a source of power and fulfilling itself in technology, which falls within the domain of technonature.

In this sense, it is immaterial that scientists are defined as all researchers recorded or affiliated to specialised societies rather than in the narrow sense of those among them who move in the corridors of power and carry more direct weight in the definition of policies. To take the example of the United States, a narrow definition considerably reduces the number of the 'ruling élite', since the estimates are never more than a thousand at Federal level and sometimes two hundred if one takes only the most influential figures.[61] But this active fraction of the scien-

61. See Christopher Wright, 'Establishment of Science Affairs', in *Scientists and National Policy-Making*, 273. In the smaller countries the group of influential counsellors does not amount to fifty and is often much less, whence

tific establishment, which holds a key position in the centres of decision, cannot be separated from the larger group of scientists who take
part as such in public affairs, or, finally, from the whole body of researchers whose work both influences decisions and is affected by them.
In this respect, there is nothing specific about the American situation,
which is reproduced in more or less similar forms and institutions in
all countries in which science is recognised by the state as a means of
attaining its general policy goals. It may be as well to recall that we
use the term politics in the wide sense of the relation to the government of the commonwealth, and that in this context no empirical
distinction between the 'active' élite and the 'passive' mass of the scientific community would in any way change the substance of the
question. 'All politics,' said Valéry, 'is based on the indifference of most
of those concerned, otherwise all politics would be impossible.'[62]

There is indeed no question of determining whether there is more or
less partisan behaviour in the group of scientists than in any other
group, more or less allegiance to political convictions and activities,
more or less cohesion and commitment to the right or left wing; from
this angle the scientist is not set apart as a subject of politics, and his
participation or non-participation raises no specific problem.[63] Neither
is there any need to imagine a plot, a conspiracy between scientists and
the 'devilish' representatives of power, the army or industry interested
in military programmes, to account for the political weight now wielded
by the scientific establishment. Technonature, it is true, may be the
scene of a trial of strength of this kind, in which the scientific lobby, in
association with other military and industrial pressure groups, confounds the defence of its own interests with the public good, but it is
not confined to such a trial of strength. The heart of the matter is the
recognition that there are very few scientific questions with which the
political power has to deal that can be divorced from their non-techni-

---

arises a real problem of renovation in the field of scientific judgment even
more than that of political decision. In general, it is well known that the number of 'active' directing élite (or 'technocrats *en puissance*', in the words of J.
Meynaud) is a narrow category rather than a wide one (see J. Meynaud, *La
Technocratie*, 47–54).

62. Paul Valéry, *Regards sur le monde actuel* (1968 edition), Gallimard,
Paris, 59.

63. Empirical inquiries, moreover, do not teach much more than might have
been expected. Thus David Nichols, following Anne Roe, shows that the
political opinions of scientists range 'from the extreme right to the extreme left,
with a majority of liberals'. (Nichols, *The Political Attitudes of a Scientific
Elite*, 192, and Roe, *The Making of a Scientist*, 228.)

cal aspects and premises. This recognition applies, in a sense, to all technical questions, but in the case of science the discrepancy between the protestation of neutrality and the reality of commitment is clearer in two ways: first, the questions which science puts to power or power puts to science are such that they involve the whole fate of societies; and second, the weight of non-technical considerations is presumed to be less in this context on the ground that the scientific spirit, methods and values are deemed to be more proof against the subjective and partisan pressures of the political arena.

Technonature is precisely the area in which the scruples and subtleties involved in the ideal separation between science and politics are cast aside. Not that the sacred and the profane are condemned to live together in wedded bliss—far from it; in the alliance between knowledge and power the stakes are not equal and misunderstandings abound. But there is no choice except to act as though the language, the needs, the interests of the one were compatible with those of the other. The scientific and technical élite has not taken over power and does not even think of doing so, but it takes advantage of it; power has not enslaved science and is very careful not to do so, but it subordinates the progress of science to the pursuit of its own ends. Technonature is not neutral, even if science still professes to be so; on the contrary, it consolidates the end of the innocence of the scientist regarded as detached from the political project which he serves and of which he makes use. The savant does not take the place of the chief or the prophet, he is at their service without having to feel that he is renouncing his vocation; in working on the object of his vocation he is at the same time doing the work of power; he may assert the scientific character of his work while at the same time integrating it in the ideology of a cause.

The fact is that he is really no longer a savant in the sense that he is in the service of truth and that, in technonature, the issue of science ceases to be that of truth: *technonature is the area in which science is revealed as a technique and the scientist as a technician of a knowledge which is significant as an instrument rather than in its relation to truth.* This area is ambivalent through and through, since it embodies, as it were, the vestiges of a dead culture, in which the goals of science were not identified with the goals of power and whose almost forgotten memory, like an ill-healed scar, is reflected in an uneasy conscience. This is no doubt what Oppenheimer had in mind when he spoke of the consequences for physicists of the perfecting of the atom bomb. The lost innocence of the physicists merely inaugurated a general situation for science as a whole, committed to politics, caught in the pitfall of the 'ethics of responsibility' even in its most abstract workings:

> In a sort of brutal significance, which no vulgarity, no pleasantry, no exaggeration can wholly abolish, the physicists have known sin, and that is a knowledge which they can never lose.[64]

64. J. R. Oppenheimer, 'Physics in the Contemporary World', *Bulletin of the Atomic Scientists*, 4, no. 3 (March 1948), 66.

CHAPTER 8

The Scientist and the Problems
of Power:
The Pitfalls of Responsibility

When the paths of science opened the way to the paths of philosophy, the savant's attitude towards the commonwealth differed little from that of the philosopher; both equally responsible towards truth, it was primarily by serving truth that they served the commonwealth—a professional duty which was all the more obvious since the savant himself, like the philosopher, distinguished himself from the technician and from the politician, both alike too eager to attain immediate ends to challenge the significance of their means. The savant was no more of a sophist than the philosopher in that, as guardian and servant of the truth, he could not maintain any thesis indiscriminately; only a science which had nothing to do with techniques could transcend the knowledge on which techniques are based.

The ambivalence of science in its relations with the commonwealth begins only from the moment when scientific activity in its turn comes in as a technique, the source at once of new knowledge and of new power, measured no longer by truth but by services rendered in attaining ends alien or indifferent to truth. For all that the scientist can once again be contrasted with the advocate, who may well forget his zeal for justice in serving the interests of his client, unlike the scientist, who knows no higher loyalty than loyalty to his professional code,[1] the fact remains that this professional duty is no protection against the conflict of opposing duties arising out of that very profession.

In Goethe's tragedy Faust sells his soul to the devil as the price of eternity; in technonature the scientist yields to the will of power in order

1. See R. Gilpin, 'Civil Service Relations in the United States: The Example of the President's Science Advisory Committee', stencilled paper presented to the convention of the American Political Science Association, Saint-Louis, Missouri (8 September 1961), 8.

to be able to ply his trade. In spite of the difference in the stakes, the process of alienation is the same; the scientist, like Faust, is merely a plaything in the hands of his partner. His goal is the advancement of knowledge in the hope that the expectation of finding a solution to the problems of science will justify the allocation of resources to finding such a solution. In the eyes of the researcher all research is good in itself and for itself, and its immediate utility is no measure of the support it deserves; its values are eternal. It has no terminal point in time, and for that very reason, the scientific process has no terminal point in space: the same tradition which asserts that truth has no frontiers lays down that the generation of new knowledge is always and at every moment good in itself, the universality of the scientific process extending into an extratemporal, non-historical legitimacy, the mark of an eminently autonomous social institution, with no other end except itself or in itself.

The state, on the other hand, has to choose, and chooses, if only because it can never have enough resources to subsidise all the activities for which it is responsible, and because among scientific activities there are some which hold out promises or offer results of special value to the state. Science is only one instrument among others in the arsenal available to it to achieve its ends; utility is the measuring rod of its interest in scientific research. 'The tempest is unleashed on high,' says Mephistopheles; 'you fancy you are the driver, but you are only the driven'.[2] The scientist fancies that, in the name of science, he influences the goals of power, but it is power which determines the goals of the research system at national level.

Science experienced as a technique

The scientist's professional duty towards the truth cannot blind him to the servitudes of organisation and even of bureaucratisation to which he is exposed by the dependence of research on the industrial system in general. Neither can he disregard the questions which the commonwealth must, or thinks it must, take into account, according to the circumstances, particularly since the content and results of his activities are not without influence on the way in which it formulates them and experiences them. This was recalled by the United States Congressional report already referred to, in language which, in other times, would have been enough to condemn the illegitimate intrusion of state reason:

2. Goethe, *Faust*, part 2.

Political decisions and policies are sometimes found necessary to mediate, postpone, or circumvent the effects of harsh and arbitrary findings of science that impose unacceptable obligations or conditions on the electorate.[3]

This is, however, a cautious formula in so far as it is levelled at the effects of science rather than at its goals or content, but which nevertheless suggests that science ought to compound with truths which are not its own.

In fact, whether he means to or not, the scientist takes sides as soon as he tries to carry out a programme which depends on state support, and at the same time he leaves the sanctuary of the serene certainties of knowledge to confront the uncertainties of action. In this sense, the philosopher is far more independent of power than the scientist has now become; he can philosophise without the state, and the fruit of his thoughts, whatever ideological use may be made of it, does not find a place in the state's arsenal of economic and technical instruments. In science, it is only the most abstract of disciplines which still escape this commitment in so far as they need neither equipment or teams. But if politics affects everything in science which is not pure theory, even 'pure' mathematics falls under its sway from the moment it has to rely on the technical infrastructure of computers.

It is perfectly true that there are few scientists who are conscious that their work is compromising the moral neutrality of science, but shutting your eyes to a problem, deliberately or not, has never proved that the problem does not exist. At a deeper level, the whole idea of the practice of research indifferent to the ends it serves may seem the nadir of the technique to which science is tending to reduce itself, in the sense in which Descartes regarded indifference as the nadir of freedom

3. *Technical Information for Congress*, report by the Subcommittee on Science, Research and Development of the Committee for Science and Astronautics, Chamber of Representatives, 91st Congress, First Session, Washington (25 April 1969), 18. The report is here dealing with the AD-X2 affair, a product of which the inventor wrongly claimed that it prolonged the life of accumulators, and on the efficiency of which the laboratories of the National Bureau of Standards, directed by Dr Astin, expressed the gravest doubts. Under pressure from a lobby, Dr Astin was forced to resign in 1953; scientific circles reacted by alerting the mass press, which spoke of a 'political aggression on science', and Dr Astin was reinstated after a long inquiry and a number of debates in Congress. For some American scientists the AD-X2 affair has retained a symbolic value as great as that of the Oppenheimer case, in spite of the disproportion between them; see even more recently, Eugene Rabinowitch, 'Political Criteria for Non-Political Jobs?', *Bulletin of the Atomic Scientists*, 25, no. 6 (June 1969), 2–3.

'which discloses a defect of knowledge rather than a perfection of will'.[4]
If the instrumental character of science is unequivocally asserted in this
claim to neutrality, it nevertheless conceals a sort of blindness, or, as has
been said, amnesia[5] towards the functions and consequences of scientific
research in modern societies. And it may well be that the men whom
C. P. Snow calls 'scientist-gadgeteers',[6] absorbed in their research with
the sole concern of achieving their project and demonstrating their
know-how, today form the majority of the 'scientific community'; that
does not thereby make them the artificers of an age in which science
would at last be freed from its political and moral implications, instead
of the blind operatives of a research organisation subordinate, like
everything else, to the production machine.

An age of science which might be called classical, in the sense that
its values depended solely on its relation to truth, closed when it became able to fulfil its promises of rapid application. The unwritten
code of social supply and demand, impossible to reduce to formal
terms, equally ambiguous in its intentions and its practice, covered the
first requirements of research which was not forced to compound with
any values except its own. The neutralist procedure of science disclaims any responsibility for the use which is made of research results,
but it is undoubtedly the expectation of these results which gives it
social legitimacy; in the name of its distinterested ends, it challenges all
servitude to the economic exigences of the modern industrial system,
but it is undoubtedly because it fulfils itself in the short term as a technique that it is accorded priority in public expenditure; it professes not
to fulfil any political function in decisions which affect the fate of
nations, but it is undoubtedly because its objectives as well as its results
form part of the setting for the political stage that it can make its voice
heard there, if only to proclaim its right to state support. The manipulation of natural forces in the light of political decisions, technonature is
completely defined by the instrumentality of the industrial system
which assigns its ends to scientific and technical research. The men who
manipulate nature can do so only at the cost of being manipulated themselves; it is an age of 'wild science' if indeed the work of reason is embodied there in values and results which the classical age would have
repudiated.

4. Descartes, *Méditation quatrième*, Editions de la Pléiade, Gallimard, Paris
(1952), 305.
5. See David Riesman, 'Private People and Public Policy', *Bulletin of the
Atomic Scientists*, 25, no. 5 (May 1959), 205–6.
6. C. P. Snow, *Science and Government*, Mentor Books, New York (1962), 64.

Autres temps, autres moeurs; it is no accident that a whole generation of scientists speak of their experience before the second world war as 'the beautiful years', the twilight of the epoch in which science was experienced not only as an intellectual adventure but as a clearing-house of values.[7] For atomic physicists in particular, Göttingen, Copenhagen, Cambridge, were sanctuaries both of research and of a culture which did not disdain to add certain moral—or simply human—qualities to care for truth. It is enough to read *The Double Helix* to see how far scientific circles and ways have travelled from the idyllic years of the 1920s, even in biology, a discipline which has nevertheless not sustained the shock of history in the same way as physics. *The Double Helix* is the first full story of the stages and circumstances surrounding the birth of a major discovery, namely the structure of genetic material, the foundation of molecular biology.[8] The joint author with Francis Crick of this discovery, which brought him the Nobel prize for medicine while he was still under thirty, Watson shows himself so irreverent towards all the accepted ideas of the sincerity, disinterestedness and noble feelings of the savant at grips with his all-absorbing passion for research, that he gives us a description of the practice of science closer to *The Threepenny Opera* than to Pasteur's idealisations. Has he blown the gaff by demythologising the pretensions of the university laboratory to be a sanctuary of purity and by describing the competition between researchers as a ruthless fight, with no holds barred, including betrayal of friendships, espionage and filching papers? Or should his revelations be put down to the brashness of an American Rastignac of science, invoking his destiny from the plains of the Mid-West: 'Look out, Stockholm, here I come!'

Not only does this testimony reveal an age of science in which the researcher, however full of genius (and often of humour), does not refrain from writing like a valet to whom no man is a hero, but above all it seems to show that the greater the number of researchers, the

7. See 'The Beautiful Years', chapter 2 of Robert Jungk's book on the history of the first atom bombs, *Brighter than a Thousand Suns*, or the testimony by C. P. Snow: 'I am going to irritate you—just as Talleyrand irritated his juniors—by saying that unless one was on the scene before 1933, one hasn't known the sweetness of the scientific life. . . . The atmosphere of the scientific twenties was filled with an air of benevolence and magnanimity, which transcended the people who lived in it.' ('The Moral Un-Neutrality of Science' in *The New Scientist: Essays on the Methods and Values of Modern Science* (ed. Obler and Estrin), Doubleday–Anchor, New York (1962), 134.)

8. James D. Watson, *The Double Helix*, Weidenfeld and Nicolson, London (1968; reissued by Penguin, Harmondsworth, 1970).

more scientific competition tends to discard all rules.[9] The moral values to which science formerly appealed are losing their sanctity and their quality to the point of seeming as meaningless as consumer goods; the world which has been disenchanted by science is indeed that of a science which is itself disenchanted, that is to say, of skills, knowledge and procedures which, general and abstract as they are, are nevertheless experienced as merely one technique among others.

Technonature and alienation

Technonature, in which the interests and attitudes of scientists are so closely associated with power that they seem to merge into the the attitudes and interests of politicians, is the favoured spot of alienated science. But we still have to take the measure of this alienation and identify its specific character. The concept of alienation, as has so often been said, is equivocal in that it may apply to anything and everything, since it primarily designates the situation in which an individual or a group finds itself unable wholeheartedly to accept that situation itself. Perhaps it is part of the essence of human nature to feel a remoteness of self from self, from other people or from the world which stigmatises an experience in life as one which is suffered rather than being deliberately chosen. In shifting the analysis from the religious or metaphysical plane of human nature to the plane of human labour, Marx has nevertheless failed to liberate the concept of alienation from its origins or from its religious or metaphysical overtones—and for good reason: alienation presupposes an essence in man in the light of which he cannot be deemed to be inherently alienated and by reference to which he can conquer his own feeling of alienation. The specific character of labour is to create works which have a continued existence outside their creators, an inevitable exteriorisation which, in the natural kingdom, is what most distinguishes man from beast. This is the theme of Hegel which Marx has developed by setting it in the context of the history, not of consciousness, but of the real and concrete world, the world of economic determinations which result in the prolongation of exteriorisation, or more accurately in its conversion into alienation, in

9. See the comments Watson's book has aroused among the sociologists who, on the basis of Durkheim's analysis of the division of labour, conclude that competition among researchers is governed by their density, the source of a bitter struggle for survival in Darwin's sense (Gerard Lemaine, Benjamin Matalon and Bernard Provansal, 'La lutte pour la vie dans la cité scientifique', *Revue française de sociologie*, CNRS, Paris, 10, no. 2 (April–June 1969), 139–64).

the sense of a relation of strangeness (*Entfremdung*) between the work and its creator, a divorce in the historical and economic experience of labour between individuals and groups and their social being. It is immaterial here that, between the 1844 *Manuscripts* and *Capital*, the principal expression of alienation changed its form and meaning for Marx, and that the repression and dispossession of which labour is the object and which fall upon all human works as a kind of ontological fatality find their key in a given type of relations of production and social organisation; the idea of alienation, in the sense of a relation of strangeness between the worker and the fruit of his work, always refers back to an essence of man which his existence is called upon to fulfil in unalienated work.

In a sense, the nature of scientific work, mainly because of the language and channels of communication through which it is expressed, already makes the scientist a stranger to other men, whether they be non-scientists or even fellow scientists whose discipline is antipodal to his own. As a result of the growing complexity and specialisation of disciplines and research sectors, there is indeed, as Oppenheimer never ceased saying, 'a growing alienation of science from the common understanding of men'.[10] This linguistic level of strangeness, as Hannah Arendt has stressed, is eminently political; scientists move in a world in which language has lost its power as a link between the experience and knowledge of all men, where the action of one group cannot follow the thought of the others.[11] The inexpressible element in scientific truths forbids us to imagine any coincidence, any immediate reconciliation between the researcher and his environment; in this sense, alienation is separation, the distance which a language and a culture set between those who have access to them and the rest of mankind, in the same way—though without the pathology, except perhaps on the part of the spectators—that madness alienates a man from other men.

This, in effect, is the origin of the 'aura' which the scientist has for the general public in modern societies, akin to the fascination which the

10. J. R. Oppenheimer, 'Science et culture', *Les Etudes philosophiques*, no. 4, P.U.F., Paris (October–December 1964), 533; see also his address at the *Colloques de Rheinfelden*, 'Spécialisation et discours commun', Calmann-Lévy, Paris (1960), 261–73.

11. 'Wherever the relevance of speech is at stake, matters become political by definition, for speech is what makes man a political being. If we would follow the advice, so frequently urged upon us, to adjust our cultural attitudes to the present status of scientific achievement, we would in all earnest adopt a way of life in which speech is no longer meaningful.' (Hannah Arendt, *The Human Condition*, University of Chicago Press, Chicago (1958), 3–4.)

fool had for primitive or medieval societies, set apart from the body politic by his unreason, but at the same time the bearer of words, visions and portents which the man of sound mind, the common run of mortals, heeds as a message transmitted from beyond the bounds of reason. There is something extra-lucid in the image of the scientist's authority which the general public and the popular press create for themselves, as though the void created by incomprehension of the impenetrable world of science could be filled only by the affirmation of an element of mystery or magic. Many Nobel prizewinners have been astonished—while others have smugly accepted it—at the authority they have suddenly acquired in the eyes of the non-scientific world and the newsmen, and which has invested them overnight with oracular powers on anything and everything. We may recall Merleau-Ponty's brilliant pages on Einstein, a 'classical spirit' if ever there was one, who was nevertheless 'a fertile parent of unreason';[12] the cost of this alienation of the scientist is that his genius is immediately reputed to be miracle-working.

But we can also speak of the scientist's alienation in another sense, and it is this sense which raises the question in what respect it defines a new and specific relation of strangeness between man and his work. Marxism would say that the scientist is in the same position as the worker, since the scientist's work, no matter how different its product, its working conditions and the status it confers, is merely a particular instance of the general situation of which the worker is the model. The 'scientific worker', however subordinate his function in the laboratory, has nevertheless little in common with the manufacturing operative who is the inspiration of Marx's thought. He is by definition a 'white collar worker' and, more than that, a member of the 'intelligentsia', a specialist marked out by his university education, assured of 'executive' privileges in an industrial undertaking and of the prestige, if not of the material advantages, of a university profession in contrast with the 'manual workers'. But it is true that the modern research system, like

12. M. Merleau-Ponty, *Signs*, 'Einstein and the Crisis of Reason' (trans. Richard McCleary), Northwestern University Press, Evanston, Illinois (1964), 195. ('A science which confuses all the patent facts of common sense, and is at the same time capable of changing the world, inevitably arouses a sort of superstition even among the most cultivated witnesses. Einstein protests that he is not a god, that these immoderate praises are not addressed to him but "to my mythical homonym who makes my life singularly difficult". No one believes him, but rather his simplicity further magnifies the legend. Since he is so astonished at his glory, and it means so little to him, it must be because his genius is not wholly of his own making. Einstein is more the consecrated place, the tabernacle of some supernatural operation.' *Ibid.*, 193.)

the whole industrial system, lends itself to the division of labour and to bureaucratisation and therefore includes, as well as its captains of renown and prestigious status, its anonymous army of poverty-stricken rank and file.[13] In the extreme case, since the key to the alienation of these anonymous or 'exploited' 'scientific workers', many of whom are in any event strangers to the functions or goals of the institutions which are their paymasters, is to be found in the nature of the régime, itself defined by the relations of production and the type of property, scientific work could be said to be—and is said to be—reconciled with itself, not reified, not alienated from the régime under which it is carried out.

This dispossession of the scientist in favour of ends which are alien to him—technology, power, profit, prestige, war, and so on—is, however, not peculiar to capitalist societies. For Marx, economic alienation, which is that of real life, in contrast with religious alienation, whose theatre is the conscience, arises not out of capitalism, but out of the division of labour. The capitalist stage of history merely brought to a climax the 'opposition between man and nature' which springs from the division of labour. Alienation is thus caused by a twofold compulsion, the compulsion to follow a form of activity determined 'naturally' by the blind pressure of nature and not 'freely' by the conscious and deliberate plans of the community, and the compulsion of social organisation based on industry which constrains man to 'make his essence into a simple means of subsistence';[14] the division of labour divides the existence of the worker, capitalism expropriates him of his essence.

It is not surprising to find the same reasoning in Marxist comments on scientific work, but this reasoning, which tends to confuse the fate of science with the fate of society in the light of the way in which society conditions science, always comes back to asserting the existence, or possibility, of two types of science, one tending towards capitalism (by expropriating from the researcher his essence), the other towards communism (by reconciling the researcher with himself). Thus, speaking of the 'fraudulent conversion' of science,[15] Frédéric Joliot-Curie drew a twofold distinction, first between science and its applications,

13. On the considerable differences in status, power of initiative, possibilities and personal satisfactions among American researchers in particular, see Spencer Klaw's series of interviews in *The New Brahmins*, Morrow and Co., New York (1968).

14. Karl Marx, *Outline of a Critique of Political Economy*.

15. F. Joliot-Curie, 'Introduction' for the World Federation of Scientific Workers, cited by P. Biguard, *Frédéric Joliot-Curie et l'Energie atomique*, Paris (1961), 127: 'I think it proper, before passing judgment, to distinguish pure scientific knowledge from the use made of it, or in short, to distinguish

and second between science practised in a capitalist world and science
practised (or as it would be practised) differently in a socialist world.
Even in a Marxist context, the idea of science as an intellectual adventure separate from action preserves the attraction of an ontological
category, an essence which cannot be reduced to the movements of history though it may suffer from being determined by history. And it is
this essence of science which warrants the hope of the practice of scientific research liberated from its alienation by virtue of a different economic and social system.

It is only one step from this conception to the utopia of Marcuse.
Marcuse, who has clearly seen that '. . . the process of technological
rationality is a political process',[16] and who is far from wrong in saying
that '. . . science has projected and promoted a universe in which the
domination of nature has remained linked to the domination of
man . . .',[17] imagines scientific reason as so thoroughly steeped in
society that any change in the orientations of society must mean a
change in the whole *structure* of science.

> Its hypotheses [i.e., the hypotheses of science], without losing their rational
> character, would develop in an essentially different experimental context
> (that of a pacified world); consequently, science would arrive at essentially
> different concepts of nature and establish essentially different facts.[18]

Thus transposed to the sphere of science, the Marxist analysis of labour
in general, referring to the essence of human nature, amounts to demonstrating that the dispossession and alienation of scientific work express
the dispossession and alienation of *an* essence of science.

But it is exactly this conception of science in the terms of an ontological category which may be suspected of being a myth. The alienation of the scientist may well be the result, like all alienation in the
Marxist sense, of fragmented labour and social organisation, but it
nevertheless retains its own special significance and form in the light
of the idea of science to which it refers, that is to say, the memory of a
'disinterested' science which does not directly affect social practice and
is not directly affected by it. The pacified world of which Marcuse
dreams might perhaps change the status of the researcher and the social
use made of his discoveries, but the concepts of nature forged by
science would not therefore be essentially different. It is not necessary

---

between thought and action in science. . . .' Scientists 'have no desire to be
the accomplices of those who, by faulty social organisation, are allowed to
exploit the results of their work for selfish and evil ends.' (*Ibid.*)

16. H. Marcuse, *One-Dimensional Man, op. cit.*

17. *Ibid.*   18. *Ibid.*

to inveigh against the social organisation or the political régime to see that modern science has no essence independent of its instrumentality, even if its underlying philosophy is not confused with utilitarianism; scientific knowledge is action, it cannot be separated from its applications, everywhere it takes the form of a close union of theory and practice, everywhere it finds its social legitimacy in the proofs it gives of its utility. It is this union of theory and practice which makes the 'fundamentalist' researcher feel that his work is alien to the use which is made of it, as though this union were not implicit from the outset in the whole purpose of his activity as a researcher—as though the quest for new knowledge could be wholly dissociated from the practical applications in which it finds fulfilment.

Here, in line with the thought of Hegel and Marx, alienation certainly involves an essence, and it is the relation of this essence to social practice which reveals science betrayed, its meaning falsified by the goals it serves or the use made of its results, research dispossessed of its character as a vocation for the benefit of ends which are alien to it. This is expressed, for example, in the report of a committee of the American Association for the Advancement of Science, which is so revealing that it is worth quoting at length:

> The ultimate source of the strength of science will not be found in its impressive products or in its powerful instruments. It will be found in the minds of the scientists, and in the system of discourse which scientists have developed in order to describe what they know and to perfect their understanding of what they have learned. It is these internal factors— the methods, procedures and processes which scientists use to discover and to discuss the properties of the natural world—which have given science its great success. We shall refer to these processes and to the organisation of science on which they depend as *the integrity of science.* . . . On the integrity of science depends our understanding of the enormous powers which science has placed at the disposal of society. On this understanding and therefore, ultimately, on the integrity of science, depend the welfare and safety of mankind. . . . Under the pressure of insistent demands, there have been serious erosions in the integrity of science. . . . The decision to land a man on the moon was hardly a fortuitous outcome of the search for knowledge. It was rather a decision, largely on political grounds, consciously to develop the basic and applied science necessary to achieve this particular technological accomplishment. . . .[19]

While recognising that the pressure of social demands has considerably reduced the time-lag between discovery and application, the

19. *The Integrity of Science,* report by the committee on science in the promotion of human well-being of the American Association for the Advancement of Science, published in *American Scientist,* Washington (June 1965).

report criticises large-scale technological operations (especially 'Project Starfish', a very high altitude atomic explosion effected in 1962, the particles from which, among other effects, might prejudice for decades the study of the natural radiations of the Van Allen belts) and denounces the threat of unforeseeable consequences involved in any hasty experiments or exploits. What is interesting in this passage is not so much its approach to the effects of technology as its underlying idea of science as dissociated not only from the technological developments which it originates or with which it is associated, but also from everything external to its own life, its development and its internal exigencies, the expression of a sort of ontological virtue of which scientists are trustees on condition that they do not surrender to the siren voices of history—will to power, domination, prestige, and so forth—and by means of which they stand surety for 'the welfare and safety of mankind'. From erosion of the integrity of science follows distortion of its true meaning in contact with history 'full of sound and fury' which is not its own; scientific knowledge is conceived as 'pure' knowledge with no other inherent effect except that of adding to its own body of thought, and obeys only the injunctions of its own internal needs, taking its stand at once on the metaphysical plane of essence and on the moral plane of integrity. And it is in so far as it meets social demands (which it nevertheless itself helps to create) that it feels alien to itself, projected into an adventure in which it no longer recognises itself, committed to enterprises whose means and ends it repudiates.

In this sense, scientific research is inevitably alienated by the mere fact that modern science has ceased to be pure theory, that is to say, an end in itself. The new knowledge finds its end not in the knowledge which is its aim, but in the object which it produces, be it only by way of information, as one means among others. In fulfilling itself as a technique, science has lost its sense of research which is its own proper end and has entered into the cycle common to all production in which each end may, in another context, serve as a means. Clothed in the instrumental character of knowledge as a producer of objects, it is no longer knowledge 'in itself and for itself' but power, action, manipulation for ends other than itself, and dispossessed of its vaunted relation to the truth, betrayed, alienated in the relation of utility where all values are reduced to the value of usefulness and where every end is transformed into a means to further ends. If the essence of science is to be an end in itself, no social organisation could ever boast of the privilege of not betraying *that* essence.

The source of this alienation is to be found in the whole purpose of modern science, in its reversal of theory and action, much more than in the social structures which have enabled it and are enabling it to attain that purpose. Ever since Galileo, access to the knowledge of nature, the means of constraining it to deliver up its secrets, the royal road to scientific truth, has lain by way of action, not contemplation. *The process of experimentation which invades the sciences not only leads to treating nature as an instrument, it also leads to knowledge itself becoming an instrument.* As Husserl has so clearly shown, the epistemological and philosophical reversal which goes back to the mathematisation of nature culminates in the growing technicality of knowledge:

> Galileo . . . is at once a discovering and a concealing genius. He discovers mathematical nature, the methodical idea. . . .[20]

He was the inventor who turned the work of thought into a mechanical operation. This work of thought, says Husserl, is 'in no way depreciated' by the fact that it was explained as a *techne*; it is in fact 'astonishing to the highest degree'. Galileo clothed the concrete presence of the real world in intelligible ideas; rendered accessible to calculation and prediction, this world, at the same stroke, lent itself to the intervention of rationality directed towards action. Knowledge was not disqualified by its metamorphosis into power—quite the reverse; but, entombing its essence as *theoria* under the growing edifice of its technical fulfilments to the point of becoming merged in them, at the same time it lost the sense of its own end. And it was in this capacity of a 'concealing genius' that Galileo embodied the alienation of modern science fulfilled as technique, an alienation which flowed from its operational character whereby it turned its back on the purpose of 'wisdom' and humanism of ancient science. The same process which replaced nature perceived by the senses with nature constructed by the theoretical and practical apparatus of science has replaced knowledge conceived as an end in itself with knowledge fulfilled as one means among others; the science which manipulates nature is itself doomed to be manipulated in return.

To this inevitable alienation, inherent in the whole definition of modern science, technonature adds another; not only is scientific knowledge no longer an end in itself, but its fate tends to become identified with that of power because its interests tend to merge with those of the state. The 'work of thought which is purely and simply mechanical'

20. E. Husserl, 'The crisis of European science and transcendental phenomenology', *op. cit.*

proves to be a mere tool in the service of power, and the problem it faces is no longer solely that of the epistemological kinship between theory and practice, but the political relation between science and the authority which is its paymaster. In the optimistic outlook of the eighteenth and even of the nineteenth century, when science was still far from fulfilling its promises of application, scientists hoped for state support without having to blend into its institutions; if they could tender their services without danger of losing control of their own affairs, it was because scientific work was still at the cottage industry stage. The industrialisation of research, which began to emerge with the dawn of the twentieth century and which was speeded up by the second world war, changed the terms of the relationship; the scientific enterprise now accepts its dependence on political decisions, and the more it can fulfil its promises of applications, that is to say, the more it holds itself out and fulfils itself as a technique, the more it depends on the options of the state. It is in this sense that technonature consolidates a new form of alienation of the scientist; subservient to the power whose decisions he affects, he invokes a conception of science which would enable him to bring pressure to bear on it, but the means which he has of exercising control have no more weight than that of one form of expertise among others. But the rationality which has enabled science to win its spurs as a technique is no guarantee against the political ultilisation of that technique; it is by his belief, with the best faith in the world, in the existence of that guarantee, that the scientist has quite simply allowed himself to become a plaything.

The significance of the Oppenheimer case

Once again, it is the story of the first atom bombs which provides the most striking example of this process of alienation, in which the scientist, creating a new political problem by his research results, loses all initiative once the results are established. A quarter of a century after Hiroshima, it is clear that the scientists, whatever influence they may have had on certain national options or international negotiations, have failed as they did in the beginning to give their common recognition of the implications of the new weapons for international equilibrium any weight in political decision. Niels Bohr was one of the first to campaign against the use of nuclear bombs on the grounds of the escalation involved in the mastery of atomic energy. In his memorandum addressed to Roosevelt and Churchill he showed himself particularly conscious of the consequences of scientific and technical competition:

. . . the further the exploration of the scientific problems concerned is proceeding, the clearer it becomes that no kind of customary measures will suffice for this purpose [the establishment of control] and that the terrifying prospect of a future competition between nations about a weapon of such formidable character can only be avoided through a universal agreement in true confidence.[21]

He went on to say that the Manhattan Project, 'immense as it was', had still proved far smaller than might have been expected and that the progress of the work had continually revealed new possibilities: '. . . any temporary advantage, however great, may be outweighed by a perpetual menace to human security'.[22] While suggesting that '. . . personal connexions between scientists of different nations might even offer means of establishing preliminary and unofficial contact',[23] with a view to security negotiations, Bohr was careful to make no suggestion whatever about the conduct and outcome of such negotiations. 'Of course, the responsible statesmen alone can have insight as to the actual political possibilities.'[24] Nearly a year later, the group of scientists centred on James Franck and Leo Szilard at the Chicago Metallurgical Laboratory went even further in submitting to Henry Stimson, Secretary for Defense, a report putting forward 'some suggestions as to the possible solution of these grave problems'.[25]

In the debate thus opened at the initiative of the scientists on the question whether the new weapons, by their very nature, could be used to eliminate war instead of preparing for it, neither Bohr nor the Chicago Group, nor even those scientists directly associated with the Washington decisions, carried very much weight. The interim committee set up to advise President Truman on post-war arrangements for atomic energy and on the use of the bomb against Japan disposed of the question by proclaiming that the new weapon, however powerful it might be, 'is just as *legitimate* as any other of the deadly explosives of modern war'.[26] Where most atomic scientists point to the specific char-

21. Memorandum by Niels Bohr, transmitted to Roosevelt (3 July 1944), cited in R. Jungk, *Brighter than a Thousand Suns*.
22. *Ibid.*   23. *Ibid.*   24. *Ibid.*
25. 'The Franck Report', in R. Jungk, *op. cit.*, appendix B.
26. Henry L. Stimson, 'The Decision to Use the Bomb' (February 1947), in *The Atomic Age* (ed. Grodzins and Rabinowitch), Simon and Schuster, New York (1965), 36. (My italics). The Interim Committee, under the chairmanship of the Secretary for Defense, consisted of the directors of the scientific war effort, including Vannevar Bush, Karl Compton and James Conant; it was assisted by a sub-group of four scientific advisers, Robert Oppenheimer, Arthur Compton, Ernest Lawrence and Enrico Fermi. See G. Hewlett and Oscar E. Anderson, *The New World*, Pennsylvania State University Press (1962), 344–61, 365–9.

acter of nuclear weapons to advocate a 'new' political approach to international relations, the answer of power is that the scientific revolution brought about by the atom, whatever its political impact, changes nothing in the traditional solutions to the problem of national security. Thus, the conflict between the authors of the 'Franck Report' and the members of the interim committee confirms the divorce of the 'scientific community'—national and international—from the ends to which science is the means. This divorce makes scientific competence and political competence turn their backs on each other; politics, in making use of science, surrenders nothing of its prerogatives, while science, in serving politics, conquers no privileged means of pressure. Technonature brutally teaches scientists that informing political decision is not the same thing as forming it.

When Bohr tried the same approach to Churchill that he had already tried with Roosevelt,

> . . . Churchill listened to the scientist for half an hour in silence. But at the end of that time he suddenly stood up and broke off the audience, before Bohr had finished. . . . The Prime Minister is then said to have turned to Lord Cherwell and asked, with a shake of the head, 'What is he really talking about? Politics or physics?'[27]

True or not, the story tells us a great deal about the scientist's power to influence political decision: what can he talk about, other than science, with more competence than any other man? The question, if not as old as the world, is at least as old as western tradition, since it is the pre-eminent question of philosophy, asked ever since Plato: if knowledge is a technique, there is nothing beyond technique to guide knowledge; what can the technician really talk about, except his technique? But however political his talk, it is not enough to endow him with political insight. The alliance between knowledge and power institutionalised by the second world war was not accompanied by a transfer of competences or a sharing of responsibilities; science became a political affair, but it did not thereby make the scientist a statesman. Politically committed by his relation to power, the scientist may hope to influence decision but not to absorb it. But, above all, in proclaiming that his advice is given solely on the technical level, he has no recourse

27. In R. Jungk, *op. cit.* See also a more recent interview with Dean Acheson, Secretary of State under Truman: 'I accompanied Oppie into Truman's office once. Oppie was wringing his hands and said: "I have blood on my hands". "Don't ever bring that damn fool in here again", Truman told me afterwards. "He didn't set that bomb off. I did. This kind of snivelling makes me sick".' (*New York Times*, 11 October 1969.)

against power once it chooses, in the name of imperatives and values which the researchers profess to ignore, technical solutions which they do not favour.[28]

It is on facts alone that the scientist can claim to base his expertise, on facts whose objectivity seems to him to be guaranteed by the method and the culture which he applies, and he refrains from looking beyond the facts as he finds them and formulates them within the limits of his methods and his culture, as though the scientific process, in applying itself to politics, automatically acquired the same universal power of persuasion that it has conquered and imposed within its own sphere of jurisdiction. Far from allowing itself to be reduced to these positivist terms, history knows no pure facts and no simple facts—even less than science itself, whose facts are constructed rather than established. Furthermore, it embroils the 'logic of facts' by a sort of ruse of reason, by investing with political passions the conclusions which scientists believe to be inspired by technical considerations alone, to the very point of passing off as purely technical those which are essentially inspired by political considerations.

If science is the quest for a truth, truth is no longer the sole end of science as a technique. Power, in making use of science, does not take its stand on this ground, but neither does science in so far as it is at the service of power; the area in which research is the subject of decisions or affects decisions is not bounded by the type of truth which it is the vocation of science to pursue. If a conflict arises—and there is no lack of them—technonature gives rise to a technical clash between two different conceptions of expertise. This is the whole difference between the trial of Galileo and the Oppenheimer case, in spite of the numerous points of comparison which always tempt us to draw a parallel. Both cases display a fundamental misunderstanding between knowledge and state reason (or church reason): in the first case it was a certain idea of

28. On the pseudo-objectivity of scientific expertise in political matters, see the commentary of the president of the Federation of Atomic Scientists at the height of the campaign to convert the American Senate to the idea of international control of atomic energy (1946): 'The Federation makes a point of being non-political. . . . To hell with politics. The question is: Are you pro- or anti-suicide?' See also David Lilienthal's comments on the work of the committee under his chairmanship which inspired the Baruch Plan: the committee had '. . . an opportunity to analyse what is called a political problem in a scientific spirit. . . . we started somewhat as a chemist might, tackling a technical problem: with the facts as he found them.' (Cited by R. Gilpin, *American Scientists and Nuclear Weapons Policy*, Princeton University Press (1962), 51, 58.)

science that was in issue, transcending the scientist as individual; in the second, the issue was a certain idea of the expert's function, transcending the individual as scientist. On the one hand, in spite of science, we have one citizen among others whose attitude and opinions ran counter to the intentions of the community; on the other, because of science, we have a researcher whose relation to knowledge challenges a whole doctrine of truth.

The difference illustrates the different climate of the age in the two cases; the political situation of crisis and McCarthyist reaction resulting from the cold war played a decisive part in Oppenheimer's case, whereas the political situation of the day, at least so far as we can judge from the surviving documents, seems to have played no part in Galileo's case. And once the crisis was over, a change of President in the United States was enough to restore Oppenheimer to favour,[29] whereas it took centuries for the church to reopen Galileo's file. It is true that the analogy could be carried down to details, and it could be shown that Galileo, relegated to Sienna and then to Florence, was no more prevented from devoting himself to research, publishing his results and exercising an influence than was Oppenheimer at Princeton, and each of them died without the charges or, indeed, the security measures taken against them being withdrawn. But between the objection of being 'a security risk' and the suspicion of heresy which hung over Galileo there is a whole world of difference—the difference between science on trial for its incapacity to serve the truth and the scientist challenged for his incapacity to serve the state.

Let us disregard here everything which is purely a matter of personal character, everything which may have seemed to be weakness, complacency or pure naïvety,[30] and concentrate on the significance of his

29. On the morning of 22 November 1963, the day President Kennedy was assassinated, the White House announced that he would personally present Oppenheimer with the Fermi Medal. On 2 December it was President Johnson who presented Oppenheimer with the medal in the presence of a distinguished scientific and political gathering. Oppenheimer ended his speech of thanks with the words: 'I think it is just possible, Mr President, that it has taken some charity and some courage for you to make this award today.' One Senator called the award ceremony shocking and revolting. Republican members of the Joint Committee on Atomic Energy boycotted it in a body. See Nuel Pharr Davis, *Lawrence and Oppenheimer*, Simon and Schuster, New York (1968), 354.

30. As demonstrated by the 'Chevalier–Eltenton incident', in connection with which Oppenheimer was caught out by the security staff of the Manhattan Project: 'There is no more persuasive argument as to why I did this than that I was an idiot.' (*In the Matter of J. Robert Oppenheimer: Transcript of Hearings before Personnel Security Board*, USGPO, Washington (1957), 888.)

clash with power. After all, Galileo abjured and that is not what matters, in spite of the lesson of prudence which Descartes drew from it; the scientist who, on his knees before the Cardinals, in the sackcloth of the penitent, acknowledged the suspicion of heresy and promised to denounce to the Holy Office every 'heretic or reputed heretic' he chanced to meet, did not thereby cease to disseminate his ideas.

> In short, he was not the perfect hero, not really a superman, but a superbly gifted mortal with more than his share of rare and admirable qualities, but also with his full share of human weaknesses. And when all that is said, it still has to be said that he was a great man.[31]

This tribute to Oppenheimer the man is no less applicable to Galileo the man. The two cases resemble each other in that they both transcend their victims; like every 'famous trial' in which the presumed guilt of an individual calls into question the proclaimed innocence of a society, they are of paramount importance because of the symbolic value they immediately acquired by indicting, above and beyond the weaknesses of the accused, the weaknesses of the accusers. But between the first case and the second, this symbolic value changed radically, just as the function and status of the scientist in society changed.

Galileo's case was clearly a conflict of authority in which science as such was suspected of heresy for refusing to accept any master except itself in its own field. Galileo asserted the need to dissociate the truth revealed by Holy Writ from the truth established by scientific inquiry. Far from challenging the scriptures, he had always proclaimed that scientific propositions, on condition of being scientifically demonstrated, served the best interests of faith, where his adversaries 'shielding their faulty reasoning behind the cloak of a feigned religion'[32] compromised them. The tribunal which condemned him invoked the sovereignty of religion over all questions, even questions which it could not determine, but 'the intention of the Holy Spirit is to teach us the way to heaven and not the ways of heaven',[33] and it is precisely this distinction between the sphere of faith and the sphere of science which gives substance to the indictment of his accusers—a distinction which he always asserted in the name of the tradition and interest of religion, conducting himself like a faithful child of the church rather than as a 'protester'. From this point of view, Santillana is not wrong in emphasising that instead

31. Robert Coughlan, 'The Equivocal Hero of Science, Robert Oppenheimer', *Life* (American edition) (3 March 1967), 34.
32. Galileo, Letter to the Duchess Christine de Lorraine.
33. *Ibid.*

of offering his adversaries a new metaphysics, like Descartes, he sought above all to recall them to their own .[34]

The two cases can also be assimilated by showing, as Santillana has done, how greatly the undisclosed intervention of the bureaucracy falsified the decision in both cases, the Pope's conviction being grounded on a meticulously contrived summary, stuffed with imputations no less insidious than those contained in the report drafted by the chairman of the Atomic Energy Commission against Oppenheimer.[35] But while it is true that false accusations and personal rancour played the same role in both cases, the parallel cannot be pushed too far; the points of differences are even more striking. The dispute between the church and Galileo was *metaphysical*—because it was scientific— arising out of two different concepts of the truth and of the limits of competence of *knowledge*; the scientist's function was challenged in so far as he invoked a truth which submitted to no jurisdiction except that of experiment and mathematical proof. The dispute between the Atomic Energy Commission and Oppenheimer, in contrast, was *political*—because it was technical—arising out of two different concepts of expertise and of the limits of the competence of *power*. The ground of debate was no longer that of scientific truth at odds with something alien to it, but of technical advice at odds with the political decision of which it was the ground. The misunderstanding of which Galileo was victim arose from the fact that the church could not recognise the scientist's *methodological* approach to the truth; in Oppenheimer's case the misunderstanding arose out of the fact that the state could not recognise the expert's *political* approach to the technical problem on which he was asked to advise.

Among all the reasons which led the Gray Board to declare Oppenheimer a security risk, this does not seem to have been the least decisive:

> . . . *he may have departed his role as scientific adviser to exercise highly persuasive influence in matters in which his convictions were not necessarily a reflection of technical judgment and also not necessarily related to the protection of the strongest offensive military interests of the country.*[36]

The decision to make the hydrogen bomb involved a whole series of

34. 'The celestial malice which Einstein admires in his writings was directed not against them but against the Diafoiruses of his day who cast ridicule on the past and on Aristotle himself; for, far from wishing to break with the past, he wanted to save it, and with it the civilisation and the culture on which he had been bred—everything which was compromised by his condemnation.' (Giorgio de Santillana, *Le Procès de Galilée*, French language version by Salem-Salomon, Club du Meilleur Livre, Paris (1955), 421.)   35. *Ibid.*, 6–7.

36. *In the Matter of J. Robert Oppenheimer: Texts of Principal Documents and Letters*, Washington, USGPO (1954), 19–20 (my italics).

diplomatic and military options which Oppenheimer, together with most of the scientists then associated with power, challenged. In the context of the Korean War and at a moment when no one knew whether the H-bomb was feasible, both partisans and adversaries proffered their advice as being based solely on technical considerations, or at least distinguished between technical and political considerations. On both sides, however, political or even simply subjective preconceptions inevitably came into play, without either side being willing to admit it. The result was that each side cast on the other the suspicion of insincerity or partisan passion from which it professed itself to be free.

Behind the stands taken by the scientists, the controversy was mainly a clash between the ground forces and the air forces on the 'tactical' or 'strategical' use of nuclear weapons. The conclusions of 'Project Vista', an appreciation made in 1951 of defence problems in the light of the possibilities of nuclear weapons, recommended the development of atomic tactical weapons, based on a conception of 'limited war' and negotiation with the Russians, in contrast with the doctrine of 'massive response' favoured by the Air Force. As Santillana suggests, the representatives of the Air Force played a similar part in the Oppenheimer case to that of the Jesuits in the case of Galileo. As leading partisan of the ideas of 'Project Vista', Oppenheimer became a target for Strategic Air Command, and it was his denunciation to the FBI as 'an agent of the Soviet Union' by William L. Borden, former personnel director of the Joint Atomic Energy Commission, and a declared partisan of the doctrine of strategic bombardment, which triggered off the affair.[37]

The epistemological revolution consolidated by Galileo's condemnation was in fact the introduction to a different culture and a different civilisation, in which, because scientific truth was disjoined from religious obedience, the distinction of orders of knowledge entailed a separation of powers among the competent authorities. The Oppenheimer case, on the other hand, tends to show that the conclusion of the secularisation of scientific research is to tie it so closely to the state and its options that there can be no strict line of demarcation between the powers of the competent authorities—nor, indeed, thanks to political tension, between the scientist's private life and his public life; secularisation is prolonged into functionalisation in the most bureaucratic sense of the word, the researcher being required to conduct himself, in

37. On the controversies of this period, see in particular R. Gilpin, *American Scientists and Nuclear Weapons Policy*, *op. cit.*, chapters 2, 4, and Warner Schilling, Paul Hammond and Glenn Snyder, *Politics and Defense Budgets*, Columbia University Press, New York (1962).

his capacity as expert, like an apolitical instrument in the service of power and to subordinate his personal relations to his professional responsibilities.[38]

Within the limits of his competence, he is not allowed to make a mistake; it is precisely because he invokes the scientific method that his technical judgment is deemed infallible, and if he falls into error, not only does he fail to come up to the image formed of his competence, but above all he reveals himself to be guilty, to be unworthy of his function. Oppenheimer never ceased to remind his judges that he opposed the H-bomb not because he doubted its feasibility, but because he was convinced that its construction, far from diminishing the dangers arising out of the development of nuclear weapons, would enhance them.[39] Teller's reasons for recommending a priority programme to develop the 'super' bomb were no less politically determined. Having demonstrated its feasibility, he could by the same stroke insinuate that Oppenheimer's political conviction was as faulty as his technical judgment, and therefore a conviction of guilt since it was flying in the face of history—as though, in the history of the sciences, a scientist must be convicted of incompetence because at a given moment in time he questions the possibility of solving a problem which is nevertheless subsequently solved; if he is right at the moment he formulates this opinion,

38. For example, the accusation levelled by the Commissioner of the Atomic Energy Agency, Thomas Murray: 'It will not do to plead that Dr Oppenheimer revealed no secrets to the Communists and fellow travellers with whom he chose to associate. What is incompatible with obedience to the laws of security is *the associations themselves, however innocent in fact.*' (*In the Matter of J. Robert Oppenheimer: Texts of Principal Documents and Letters,* Washington, USGPO (1954), 63; my italics.) Oppenheimer was blamed, too, for having seen, during the Los Alamos period, a former fiancée whom he knew to have been a Communist: 'You spent the night with her, didn't you? Did you think that consistent with good security?' (*Transcript of Hearing,* 154.)

39. The General Advisory Committee of the Atomic Energy Commission, under Oppenheimer's chairmanship, adopted a negative attitude towards thermonuclear weapons but was split into two groups. The minority, represented by Fermi and Rabi, stressed the ethical issues at stake. 'The fact that no limits exist to the destructiveness of this weapon makes its very existence and the knowledge of its construction a danger to humanity as a whole. It is necessarily an evil thing considered in any light. For these reasons we believe it important for the President of the United States to tell the American public and the world that we think it wrong on fundamental ethical principles to initiate the development of such a weapon.' The majority group, including Oppenheimer, laid greater stress on political issues: 'In determining not to proceed to develop the super-bomb, we see a unique opportunity of providing by example some limitations on the totality of war and thus of eliminating the fear and arousing the hope of mankind.' (*In the Matter of J. Robert Oppenheimer: Transcript of Hearing,* 79–80.)

he will not thereby be wrong after the event. But the time scale of the history of science is not the same as that of the functionalised scientist; he must be right or wrong, since his technical advice is the groundwork, here and now, of political decision. Even Oppenheimer ended by recognising this when he reduced the reasons for his conversion to the H-bomb solely to the changes in the technical conditions for its construction:

> The programme we had in 1949 was a tortured thing that you could well argue did not make a great deal of sense. *It was therefore possible to argue also that you did not want it even if you could have it. The programme in 1951 was technically so sweet that you could not argue about that.* It was purely the military, the political and the humane problem of what you were going to do about it once you had it.[40]

Thus the loyalty which the scientist by his vocation owes to the truth comes into conflict with the loyalty which by his function he owes to the state. Just as Galileo could neither understand nor admit the accusation of the Holy Office, since by invoking scientific truth and denying that he was compromising the truths of faith he disclaimed the power of the latter to limit the former, so Oppenheimer was bound to yield to the reasons of his adversaries since, in taking his stand on the ground of technical expertise, he recognised their right to arrive at a different conclusion from his own. Galileo could appeal to eternity, Oppenheimer had no recourse against history: *the scientist's vocation is alienated in his function.* The paradox, moreover, as Sanford A. Lakoff has clearly seen, is that it was Teller and not Oppenheimer who invoked the idea of science as an intellectual adventure to which power can set no limits, whatever the consequences.[41] Emphasising the disappointment of the Los Alamos scientists when they learned of the adverse report of the General Advisory Committee on the H-bomb, Teller said:

> ... the people there [at Los Alamos] were a little bit tired—at least many, particularly of the younger ones—of going ahead with minor improvements and wanted to in a sort of adventurous spirit go into a new field. ... Not only to me, but to very many others who said this to me spontaneously, the report meant this. As long as you people go ahead and make minor improvements and work very hard and diligently at it, you are doing a fine job, but if you succeed in making a really great piece of progress, then you are doing something that is immoral.[42]

40. *In the Matter of J. Robert Oppenheimer: Transcript of Hearing*, 251 (my italics).

41. S. A. Lakoff, 'The Trial of Dr Oppenheimer', in *Knowledge and Power* (ed. Lakoff), The Free Press, New York (1966), 81. See also J. Stefan Dupré and Sanford A. Lakoff, *Science and the Nation*, Prentice-Hall, Englewood Cliffs, N.J. (1962), 148–9.

42. *In the Matter of J. Robert Oppenheimer: Transcript of Hearing*, 716.

What made Dr Oppenheimer suspect in the eyes of his scientist adversaries, however says Lakoff, was not simply that he was 'wrong' in his technical judgment, but rather the inconsistency of his adherence to the code of the vocation.[43] The inconsistency at least shows that Oppenheimer resisted the alienation of technonature which Teller took for granted; for Teller, science as an instrument of power did not thereby cease to be an intellectual adventure, whereas Oppenheimer challenged the function of science as one technique among others whose ends or consequences could not be questioned. What makes the Oppenheimer affair symbolic, going beyond the classical problems of security in time of war and the climate of anxiety, and even of hysteria, of the cold war out of which it arose, is precisely this new version it provides of difficulties encountered by knowledge in its relations with power: the code of the vocation becomes embroiled, it submits so easily to the imperatives of the state, that it confounds political function and vocation, as though the latter in fulfilling itself through the former still retained its character as a vocation, whereas in fact knowledge is subordinated to power. The problem no longer lies between the way to heaven and the ways of heaven; it lies in doing the will of the state while still being faithful to a vocation. Oppenheimer's 'inconsistency' is that of the scientist who, caught in the pitfalls of responsibility, cannot find the answer to the problem as lightheartedly as Teller.

The big city and the village of knowledge

It is an oversimplification to divide scientists into those who devote themselves to their calling and those who make political pronouncements, even if only on scientific questions. From this point of view, it is clear that the authority with which they are vested by science cannot be transferred:

> Learned societies, as soon as they debate war and peace, become political and not scientific associations. Their appeals would be more convincing if they did not often display a naïvety in diplomatic matters commensurate with the authority confidently accorded to them in matters of nuclear physics.[44]

In this sense the removal of Frédéric Joliot-Curie, who committed his authority and functions at the head of the French Atomic Energy Commission to public statements in favour of the Communist Party,

43. S. A. Lakoff, 'The Trial of Dr Oppenheimer', *op. cit.*
44. R. Aron, introduction to Max Weber, *Le Savant et le Politique*, Plon, Paris (1959), 17.

has never been regarded as an example of science as such becoming the victim of politics.[45] But it is not only by committing himself as a citizen that the scientist may clash with power, it is also by assuming his responsibilities as scientist. If science is a source of power, the progress of knowledge regarded as a good in itself cannot be divorced from its consequences or from the use made of its results. The problem Oppenheimer had to face has spread beyond the man and the field of research in which it so brusquely originated; it is not only nuclear research but scientific research as a whole which raises the same questions.

For all that the menace of nuclear cataclysm might seem to be out of all proportion more dramatic—or more 'contiguous' in Hume's sense of the word and therefore more keenly felt—than that of discoveries in other fields (such as molecular biology with its possibilities of manipulating genetic capital), *it can be regarded as an exceptional case only at the cost of disregarding the continuous process of the modern research system.* Just as there are no strict frontiers between basic research and applied research, there are none between military research and civilian research. The possible revolt against the proportion of total research financing devoted to military research masks the reality of technological development which cannot be dissociated from the research effort in general. And it is the impossibility of distinguishing clearly between research which cannot be 'diverted' from its ideal ends and research whose objectives are clearly set by the paymaster institution which stops us from reducing the problem to its purely military aspects without straying into casuistry.

It is, indeed, impossible to avoid casuistry in discussing the moral problem of the scientist in his relations with power; how can he draw the line between research which remains fully under his own control and research whose use seems to him to conflict with the ends of science as he sees them? The scientist, says Snow, is not a soldier; he is not bound to blind obedience: 'Scientists have to question and if necessary to rebel.'[46] But in the first place the decision to dissent also has its social counterpart which, under certain régimes, may go further than exclusion from power and amount to the renunciation of all scientific activity and even to physical elimination:

> Only a very bold man, when he is a member of an organised society, can keep the power to say no. I tell you that, not being a very bold man or

45. See, in particular, B. Goldschmidt, *Les Rivalités atomiques, 1939–1966,* Fayard, Paris (1967), 158–88.
46. C. P. Snow, 'The Moral Un-Neutrality of Science', in *The New Scientist, op. cit.,* 136.

one who finds it congenial to stand alone, away from his colleagues. We can't expect many scientists to do it.[47]

Finally, and above all, outright rejection, whatever its value as bearing witness to the faith, does not change one jot or tittle in the world order or, specifically, in the current of research.[48] From refusal to collaborate, through conditional collaboration to wholehearted association, the scientist's attitude to power ranges through the whole spectrum of commitment to the commonwealth, and there is no answer to the question except that provided by the individual conscience.[49] The scientist, by very reason of his activities, cannot escape the antinomies of the 'ethic of responsibility' which Max Weber attributed solely to the man of action. '*Hier stehe ich, ich kann nicht anders; Gott helfe mir, amen!*': the vocation of science for truth and universality does not stop it from compelling, in its social function, choices between compromise, if not to save the commonwealth at least to ensure the progress of science, and unconditional refusal with its consequent sacrifices.

It may be said that in wartime the manifest risk and issues and collective mobilisation make the choice, if not easier, at least less doubt-ridden, while in peacetime the alliance with the state, and particularly with the military, is less obviously essential.[50] But the conditions of modern strategy and its inseparable partner technological competition, as they result from the progress of science itself, involve a mobilisation of researchers and an exploitation of research results in general which inevitably perpetuate the dilemma of responsibility: where does scien-

47. *Ibid.*, 137.
48. As witness the history of the French atom bomb, to the achievement of which, in the words of General Ailleret himself, very few scientists contributed: 'A second reason for thinking that our programme failed to develop with all possible speed is that French science did not play the part in the undertaking which, in my opinion, it should have done. . . . Our team at the Atomic Energy Commission which set about creating our nuclear armament was a team of top engineers—weapons engineers, specialised atomic officers seconded from the military staff—a team of engineers and not of scientists proper.' (General Charles Ailleret, *L'Aventure atomique française*, Grasset, Paris (1968), 396–7.)
49. On the gradation of attitudes of rejection, see J. Meynaud and B. Schroeder, *Les Savants dans la vie internationale*, Etudes de Science politique, Lausanne (1962), chapter 4, 63–75.
50. None of the scientists invited to take part in the Manhattan Project refused. Arthur H. Compton cites the example of Volney Wilson, who objected at first, saying 'It is going to be too terribly destructive. I don't want to have anything to do with it', but offered his services after Pearl Harbor with the comment, 'A lot of water has gone under the bridge since I was thinking that way'. (Arthur H. Compton, *Atomic Quest: A Personal Narrative*, Oxford University Press, New York (1956), 42.)

tific war effort begin? Refusal to cooperate is in fact no more than bearing witness to a conscience which desires to be at peace with itself, but
which can have no illusions about the efficacy of its decision. Thus,
Norbert Wiener, breaking off all collaborattion with the military in
1947, wrote in an open letter,

> I can only protest *pro forma* in refusing to give you any information con-
> cerning my past work. . . . I do not expect to publish any future work of
> mine which may do damage in the hands of irresponsible militarists.[51]

If it were expressed in the name of moral imperatives, common to
all men by the universality of the moral law, in Kant's sense, such a
refusal would call for no comment; but it is expressed in the name of
the specific imperatives of science, or at least of an idea of science
whose standards cannot claim the same universality. Reaffirming his
refusal two years later, Wiener made his position clear:

> . . . it is clear that the degradation of the position of the scientist as an
> independent worker and thinker to that of a morally irresponsible stooge
> in a science-factory has proceeded even more rapidly and devastatingly
> than I had expected. The subordination of those who ought to think to
> those who have the administrative power is ruinous for the morale of
> the scientists.[52]

The idea that the scientist's vocation demands a moral code of its own,
in line with the will of science to universality, comes up against the
impossibility of separating from research results those which will be
beneficial, and still more of foreseeing in what way a discovery regarded as an advance of reason may take shape as yet another instrument of the irrationality of history.

> Scientists should stand up and count heads, not on the question of the
> technical applications of their researches, but on their adherence to the
> moral principles which have made science possible,

says Eugene Rabinowitch, one of those who have done most to convince
his scientist colleagues of the social responsibilities they assume by the
mere fact of their research work.[53] And he proposed the introduction of
a 'Hippocratic oath' by which scientists would vow 'to work solely for
the good of mankind'. We can, of course, parade the special values of
science, those 'moral principles' which have made it possible as a pursuit

51. N. Wiener, Letter published by the *Bulletin of the Atomic Scientists*,
3, no. 1 (January 1947).
52. *Ibid.*, 4, no. 11 (November 1949).
53. E. Rabinowitch, 'Science Faces a Double Danger', *ibid.*, 9, no. 2, March
1953).

of the truth, disinterested effort to understand the mechanisms and
functioning of nature, an intellectual activity whose prime end is not
utility, but these values are set in a utilitarian context to which science
itself is by no means a stranger, and they cannot be translated into
human action by the mere fact that the scientists are unanimous in
subscribing to them more loyally than to the common values of the
community.

National or ideological loyalties, or more prosaically the demands of
a career, rarely take second place to the tribute payable to the vocation
and moral code of an activity which can be carried on, often with
genius, without its standards being treated as imperatives. The scientist
deludes himself when he relies on the neutrality of the scientific process
to disregard its social function. As Bertrand Russell wrote, he cannot
honestly say, 'My job is to provide knowledge and I am not responsible
for the way it is used.'[54] But *at the same time*, the commitments to
which he is bound by his vocation are not enough to guarantee him the
means of discharging that responsibility. The whole process by which
science is henceforth fulfilled as a technique rules out the old optimistic
eighteenth-century view of science as the automatic source of progress.
The type of reason grounded on science, and on which it relies for its
own internal progress, is not transmitted to history as a consequence of
its own successes.

What can the vocation of the scientist do in the affairs of the city, if
it merely offers power the use of one technique among others? Organised, industrialised modern society, subject to the imperatives of
production and the outbidding of competition between nations, reduces
the old antagonism between knowledge and power, placed under the
auspices of truth, to a personal choice between a vocation which cannot
be fulfilled without the support of power and the compromise of that
vocation by the very support it receives. Let the scientist but refuse to
live science as just one technique among others, let him expect *more*
from his vocation, and he will be a mere rebel in the scientific 'big city'
which diminishes science to a production activity instead of honouring
it as a free intellectual adventure.

Thus Oppenheimer described the sole refuge of the scientist as the
'university village', the haven of peace, creation and profundity for all
who reject the alienation of technonature. It is just too bad if the 'big
city' lays its hands on the discoveries of the 'village' and distorts their

54. Bertrand Russell, 'The Social Responsibilities of Scientists', in *The New
Scientist, op. cit.*, 115.

meaning; the clamour of the industrialised, utilitarian world, obsessed with power and technology, is deadened before it reaches the village of knowledge which preserves the meaning both of disinterested research and of the essential human qualities—intellectual exchange for its own sake, 'freedom' of dialogue no less than of inquiry, 'intimacy' of human contact and of conscience—in contrast with the city, 'too vast and too disordered', a prey to 'the tyranny of mass communications and professional advancement' where there is 'neither sense of community nor objective comprehension'.[55] Galileo did not despair of the world which had condemned him, since it had no jurisdiction over the combat he had joined, the real issue of which, by definition, eluded it. Oppenheimer, cold-shouldered by power, resigned himself to a society where the cards are marked and in which the scientist can do nothing except preserve his professional skill by escaping from the crowd, the tumult and the chaos which result from the irrational use of the conquests of reason.

This withdrawal of the scientist into the village is a sort of phantasm of the return of science to its maternal breast, the return of a science as innocent and pure as on the first day, that is to say, dissociated from its effects, its social function, its influence on world affairs, in a word, pure *theoria*. The purity of science, its integrity, its essential values, coincide with the universality of its processes; the *university* is the ideal scene of this coincidence, in so far as it escapes the contingent processes of history, the corrupting pressures of private interest no less than the impostures of the historic destiny of the community; it is the scene of an immediate relation between man and knowledge, just as for Rousseau the state of nature was the scene of an immediate relation between man and the natural world. And just as for Rousseau man was perverted by the mere act of yielding to his destiny, so science is lost by the mere fact of blending with history. 'The evil which comes from the outside is the passion for the outside,' says Jean Starobinski of Rousseau's genealogy of corrupted society, beginning with the end of the self-sufficiency of the natural state.[56] For Oppenheimer too, evil came from outside; it was temptation, attraction, passion for what was outside the sphere of the universality of science. For the scientific community the

55. J. R. Oppenheimer, 'Prospects in the Arts and Sciences', in *The Open Mind*, Simon and Schuster, New York (1955), 144–5. See also the same author's 'Science and Culture', *Les Etudes philosophiques*, P.U.F., Paris, no. 4 (October–December 1964), 533–5.

56. J. Starobinski, Introduction to 'Discours sur l'origine et les fondements de l'inégalité', in *Oeuvres complètes de Rousseau*, Editions de la Pléiade, Paris (1964), volume 3, lxi.

autonomy of the university village fulfils the same function of a non-historical benchmark as the self-sufficiency and solitude of the natural state fulfilled for the human community in Rousseau's thinking.

The 'passion for the outside' alienates and compromises the recognition of man by man, substituting appearance for reality in the relation of man to other men and to the world; outside the university village, there is no recognition of the scientist by his peers, the essence of science being alienated and compromised in the instrumentality of the city. A benchmark outside time, the state of nature has vanished, following a beginning which was itself outside time; it is in leaving the village that science finds its vocation for the universal immediately corrupted, just as the natural man, departing from his self-sufficiency, immediately enters upon the process of social corruption. A fatal instant, without date or place, in which science bids farewell to its essence and integrity, like Rousseau's natural man bidding farewell to his being and his goodness.

But the university village is *already* a scene of history, science is *already* distracted from its essence, just as the state of nature is *already* denatured by the history of the 'almost imperceptible progress from the beginnings'.[57] A reflexive fiction in origin and indentity, the 'pure' scientist does not exist any more than the 'noble savage', and the return to the village is as imaginary as the return to nature; everything was played out in that indeterminate and irreversible instant when knowledge entered into operative science like Rousseau's natural man entered into nascent society. The scientist is neither artist nor anchorite, but a worker whose very vocation depends on the functions society assigns to him; science is irreversibly ambivalent as an intellectual adventure and realised as a technique, the scientist irreversibly condemned to the ambiguities of action in respect of which the objectivity and universality of scientific thought guarantee him neither infallibility nor immunity. As demonstrated by the student revolt, one of whose themes, especially in the United States, is the challenge to the meaning of knowledge when it is associated with power, the 'scientific village' is more than ever besieged by history; just as the university laboratory is not outside technonature, the scientist is not outside politics. And the problem has already ceased to be whether either of them can escape; it is how far in the welter of inescapable responsibilities they can be answerable to them without completely denying themselves.

57. J.-J. Rousseau, 'Discours sur l'origine et les fondements de l'inégalité', *Oeuvres complètes, op. cit.*, vol. 3, 167.

CHAPTER 9

Science International

The ambivalence of the 'neutralist' pretensions of scientific thought—legitimate and deceptive at the same time—is nowhere more evident than in the field of international cooperation. Science is universal by nature: the truths which scientists pursue are not national truths; they are the same everywhere, and can therefore be universally recognised. International by nature, science is also international in structure: research and discovery, whatever the particular genius of individuals or peoples, take shape as a common and cumulative work, whose progress is the result of efforts by scientists all over the world, each of them adding his stone to the edifice in time and space, each depending in his own search for truth on the truths established by others. The rise of modern science ever since Galileo has been based on two principles which are the foundation of every form of cooperation in this field: cooperation between specialists, and the publication of knowledge and discoveries.

Because of its objectivity, science is deemed to be 'supracultural', to be immune from the conflicts of values to which all other expressions of culture are doomed; in the same way, its universality is alleged to confer a specific supranational character on the scientific institution and the men who are its members. The whole idea of national scientific communities is contradictory. There can be only one scientific community, which is therefore bound to be international; a single language, similar procedures, equivalent experience, common standards, are all so many characteristics which should distinguish scientific activities from all other forms of activity.[1] The idea of a scientific community is the idea of a spiritual brotherhood whose rules and bonds are born of the spirit only and therefore transcend the clashes born of nationalisms or ideologies to the point of disregarding their very existence.

1. The most idealised expression of this theme is by Michael Polanyi, 'The Republic of Science: Its Political and Economic Theory', in *Criteria for Scientific Development: Public Policy and National Goals* (a selection of articles from *Minerva*) (ed. E. Shils), Cambridge, Mass. (1968).

In this sense, 'pure' research at the same time provides all scientists with the perfect pattern of a scientific institution[2] and offers all men the most finished image of mankind without frontiers. The more researchers dedicate themselves to purely intellectual work, the more conscious they will be of belonging to the same family; the more research is oriented towards applications, the greater will be the strain on the bonds which unite this family. Science envisaged as an intellectual adventure is the foundation of an internationalism which crumbles under the pressures of national antagonisms as it turns more and more towards achieved techniques. Thus, in the hierarchy of 'pure' and 'impure', of fundamental' and 'applied', the idea of the scientific community is already restricted to those researchers whose work is essentially intellectual, and even among them to those with the greatest scientific gifts:

> The supranational character of scientific concepts and language [said Einstein] is due to the fact that they were formed by the greatest brains of all countries and all times.

To which comes the echo of Szent-Györgi's words:

> Newton is my colleague, and Galileo. A Chinese scientist is much closer to me than my own milkman.[3]

The international character of science is legitimised by intellectual agreement on the works of the intellect, but its foundations are all the firmer if these works are destined to serve mankind as a whole. This is a Cartesian theme if ever there was one, whose political extension, as demonstrated by Father Dubarle, does in fact go back to Descartes.[4] 'Reason is naturally equal in all men'; if diversity of opinion brings them into conflict, it is not because 'some are more reasonable than others', but because 'we guide our thoughts along different paths and

---

2. According to the sociologist Norman W. Storer, 'this central cadre of scientists [basic researchers] constitutes the principal reference group for most of the other groups [applied scientists and engineers]—the norms and values characterising it serve as standards for the others'. *The Social System of Science*, New York (1966), 15.

3. Albert Einstein, 'The Common Language of Science', translated from *Conceptions scientifiques, morals et sociales*, Flammarion, Paris (1952), 149; Albert Szent-Györgi, reported in *The Observer*, London (24 November 1957), and cited by Jean Meynaud and B. Schroeder, *Les Savants et la Vie internationale*, Lausanne (1962), 148.

4. D. Dubarle, O.P., 'The Proper Public of Science: Reflections on a Cartesian Theme concerning Humanity and the State as Audience of the Scientific Community', *Minerva*, 1, (summer 1963), 405–27.

do not look upon the same things'.[5] If only men will look at the same things and guide their thoughts along the same paths, that is enough for the irrefutable evidence of the truth to act as a connecting link: the language of reason sweeps away everything which, by blinding them to the fact that they belong to the same *species, accidentally* separates them.

In part six of the *Discours de la méthode*, Descartes explained how Galileo's condemnation stopped him from publishing his treatise on physics, although he thought there was nothing in it which was 'prejudicial to religion or to the state', and how nevertheless, by dint of progressing in the knowledge of physics, he decided to publish the *Discours*; he could not keep secret these general concepts he had formed about physics 'without gravely sinning against the law which commands us to ensure, as much as lies within us, the general well-being of all men'.[6] The international character of science is founded not only on its universality but also, says Father Dubarle, on the 'generosity' it displays in so far as it constitutes 'the first act of the mind which can be truly described as ecumenical'.[7] The universality of science appeals to the universal element in mankind; transcending the frontiers born of the accidents of history, science cannot associate itself with either church or state, in so far as each of them alike embodies the particularisms and divisions generated by historical contingencies.

And yet science has associated itself with the state, both because it needed the state and because it hoped to profit from the state: an evolution 'in many respects inevitable and in the nature of things'.[8] It is precisely the utility of scientific knowledge, invoked by Descartes to harness it to the common good of mankind, which subjugates its universal quality to the accidental separations of history. The rise of modern science as a collective *praxis*, based not on its value as truth but on the applications it promises, brings the twofold loyalty of the scientist, towards science and towards mankind, under the sway of the common law of national loyalties. The legitimacy of science conceived as a link among men then becomes an accepted idea specific to the ideology of science: the triumph of the operational character of modern science, as a fulfilled technique rather than an intellectual adventure, sweeps away all correspondence between the universality of its procedures and the universality of the pattern which the scientific com-

5. R. Descartes, *Discours de la Méthode*, Editions de la Pléiade, Paris (1963), part 1, 126.

6. *Ibid.*, part 6, 168. 7. Dubarle, *op, cit.*, 412. 8. *Ibid.*, 419.

S.P.—8*

munity professes to follow when it claims to be a society in its own right. This is where the mystification begins.

The scientist as a symbol of the state

> I would rather [wrote Leibniz] see the sciences flourish in Russia than see them mediocrely cultivated in Germany. The country in which science fares best will be the one which will be dearest to me, since the entire human race will benefit from it always.[9]

The society of sciences which he founded in 1700 was to labour for the honour of German science and the German language, but the national context of a scientific institution did not thereby mean the appropriation of knowledge and results which the ecumenism of science dedicates to the whole human race. Competition between researchers does not—in general—assume the unbridled form taken by the rivalries of trade, industry and politics, and the disappointment of failing to outdistance a foreign colleague is in theory no greater than if priority in discovery falls to a countryman.

And yet, if the scientific 'victories' gained by one country do not mean that the other countries are 'defeated', the international recognition and prestige attaching to prizes awarded by committee of peers are reflected from the prizewinners to their respective countries. It is, of course, science which the Nobel Foundation honours in the persons of the scientists whom it singles out, but a country is no more indifferent to piling up a big score than it is to reaping a harvest of gold at the Olympic Games. The practical imperatives of research do not give the scientific community any greater solidarity than that of any other professional group whose proceedings and interests involve international relations. The history of international commerce, it is said, discloses equally strong, or even stronger, links between financiers, cartels and trusts (as evidenced during the wars between the United States and Germany), and it would be difficult to maintain that these 'communities' carry less weight in world affairs.[10]

So long as science was defined as the theory and practice of knowledge separable from its applications, scientists could argue that their connections and their exchanges had a specific transnational character. And, in fact, for a long time international scientific relations did not depend on political intervention or commitment. This was firstly

9. Leibniz, letter to Count Golofkin (16 January 1712), in *Oeuvres* (ed. Foucher de Careil), 2nd edition, Paris (1875), vol. 7, 502–3.
10. See Arnold W. Frutkin, *International Cooperation in Space*, Prentice Hall, New Jersey (1965), 10–17.

because cooperation between scientists in different countries was limited to the exchange of correspondence, publications and visits; science was an individual affair, and neither the type of research nor the instruments it needed called for team work, still less for an international team. Secondly, and above all, science counted for little in national life; the time-lag between laboratory discoveries and their practical applications was too long, and the applications themselves too limited, for science to be recognised as a 'national asset', a source of power and independence, a factor with a two-way influence on foreign affairs.

This does not mean that, before the Atom Age, 'the sciences were never at war'. Gavin de Beer has devoted a whole book to defending and illustrating this proposition, which the experience of the French Revolution, and even the exchange of correspondence between Pasteur and his German colleagues after the Franco-Prussian War, show to be little more than a golden legend.[11] It is true that, at the height of the hostilities between England and France, Jenner was able to use his influence to help the return of scientists interned on both sides, but during the Napoleonic campaigns and those which followed them the exchange of prisoners was never limited to scientists. The case of Humphrey Davy speaking at the Institut de France during the Napoleonic Wars, receiving a prize and conducting public experiments in the hotel where he was staying, should not be allowed to mislead; this 'benevolent neutrality' on the part of countries at war was due not so much to any special respect for science as to ignorance or to the slight consequences of the services which science could render to those countries. Happy times, no doubt, when the 'brain hunt' was not yet on, and it was inconceivable to anyone that a scientist might also be a diplomat, a soldier or even a spy.[12]

11. Gavin de Beer, *The Sciences Were Never at War*, Nelson, London (1960). On the clash between certain French and German scientists, and particularly between Pasteur and Quatrefages and Virchow, see *La Revue scientifique de la France et de l'étranger* (1870, 1871 and 1873), a feast of polemics on either side where scientific considerations are couched in language worthy of the most aggressively nationalist press.

12. It is child's play for Frutkin to demonstrate that the examples which Gavin de Beer particularly stresses are far from convincing. It is true that, at a moment of grave tension between England and the United States, Benjamin Franklin gave Cook a safe-conduct, but the gesture was an empty one, as Cook was already dead. It is also true that around the same time Captain Flinders rendered assistance to Nicolas Baudin's French scientific expedition in Australian waters, but Flinders, returning to England, was arrested by the French Fleet and interned as a prisoner of war for seven years. 'It was,' says de Beer, 'one of the harsh ironies of history and a black page in the records

A common language presupposes a common understanding, but effective communication is no guarantee of cooperation, and still less of solidarity. The common language is still only a purely technical means in relation to an extrascientific end, over which it has no preferential influence. The world wars of the twentieth century have shown the limits of this confusion of concepts between communication and co-operation, between what is and what should be, by breaking the bonds between scientists, just as they divided the socialists in 1914 or the pacifists in 1939. The internationalism of science is not a sufficiently strong ideological cement to unite the international community of scientists, particularly as it it precisely since the nineteenth century that the needs of scientific research, just as much as its consequences, have made scientists more and more dependent upon states.

Just as it paves the way for science policies at national level, so the politicisation of science precipitates the intrusion of history and ideological conflicts into international scientific relations. So little do these relations escape the restraints of technonature that the free circulation of ideas, the principle and instrument of their progress, is halted by the imperatives of secrecy. On the eve of the second world war, the Harvard physicist Percy W. Bridgman stated publicly that he would in future refuse to open his laboratory or to communicate the results of his experiments to any citizen of a totalitarian state. 'No liberty for the enemies of liberty!' Transposed to science, the words of Saint-Just amount to imposing secrecy in the name of its requirements of publicity. Thus the ideology of science, conceived as the pursuit of truth, calls for its defence by arms other than ideology:

> Science has been rightly recognised as probably the one human activity which knows no nationalism. . . . But it seems to me that the possibility of an idealistic conception of the present function of science has already been destroyed, and the stark issues of self-survival are being forced upon us. . . .[13]

This call for secrecy of communications immediately met the opposition of scientists, who denied any right of politics to force its way into the laboratories:

---

of Anglo-French cultural relations.' He is bound to recognise, however, that Flinders was arrested by way of reprisals, one of the two ships in the French expedition, *Le Naturaliste*, having been seized on the homeward voyage with all its scientific cargo, in spite of the safe-conduct granted by the British (Gavin de Beer, *op. cit.*, 112–21, and Frutkin, *op. cit.*, 12–13, 25, note 1).

13. P. W. Bridgman, 'Statement', *Science*, no. 2304 (24 February 1939).

Science itself is imperilled far more than is any hateful political system when those engaged in the search for truth utilise scientific institutes, scientific laboratories or scientific journals as weapons in political warfare.[14]

Two months later, Leo Szilard, and some of the European atomic scientists who, like him, had emigrated to the United States, asked their French and British colleagues to impose self-censorship on the results of their work: the approach met with no response, particularly on the part of Joliot-Curie who turned a deaf ear for reasons of priority and national prestige rather than of conformity with the scientific imperative of the free circulation of ideas.[15] The outbreak of war was to sweep away the last scruples which these appeals for self-censorship might arouse.

The adventure of the Manhattan Project, which turned university laboratories into annexes of the arsenals, not only delivered over the scientists to the common fate of mobilised citizens, but turned them into agents of the state. Recalling how difficult it is to define the field of study constituted by international relations, Raymond Aron proposes the following images:

> Interstate relations are expressed in and by specific actions, those of individuals whom I shall call symbolically the *diplomat* and the *soldier*. Two men, and only two, no longer function as individual members but as *representatives* of the collectivities to which they belong: the diplomat in the exercise of his duties *is* the political unit in whose name he speaks; the soldier on the battlefield *is* the political unit in whose name he kills his opposite number. . . . The diplomat and the soldier *live* and *symbolise* international relations which, in so far as they are interstate relations, concern diplomacy and war.[16]

Post-war events were to show that there is a third symbolic figure alongside the diplomat and the soldier who, like them, intervenes in diplomacy and war and whose symbol is all the more closely bound up with the state since the state cannot do without him in defining and carrying out its foreign policy.

14. Douglas Johnson, *Science*, no. 2307 (17 March 1939).
15. See R. Jungk, who was given the following explanation of Joliot-Curie's refusal by one of his colleagues: 'We knew in advance that our discovery [the chain reaction of uranium] would be hailed in the press as a victory for French research, and in those days we needed publicity at any cost, if we were to obtain more generous support for our future work from the government.' (*Brighter than a Thousand Suns, op. cit.*, 179.) On the campaign for self-censorship led by Leo Szilard, see his own 'Reminiscences' in *The Intellectual Migration* (ed. D. Fleming and B. Bailyn), Cambridge, Mass. (1968), 108–18.
16. Raymond Aron, *Peace and War—A Theory of International Relations*, New York (1966), introduction, 5 (author's italics).

As scientific or military attaché where he is not strategist or ambassador, the scientist takes part in international consultation and negotiations in the same way that he is committed, from the laboratory to the battlefield, to the death of his neighbour. Interstate relations are conducted through the medium of science and with science as their subject matter: they involve science in diplomacy and war so thoroughly that there are scientists who live and symbolise them in the very exercise of their functions just like diplomats or soldiers. The specific character of the scientist whose activity, culture and values were indifferent to the world of politics, or even in revolt against it, vanishes with the specific conduct by which, in his turn, he expresses and represents the state in its relations with other states.

But here again, the Hiroshima explosion did not initiate but merely speeded up an evolution which had already been going on before the present century. It is no accident that the awakening of nationalisms coincides with the first institutionalised forms of interstate cooperation. The rise of industrialisation, the progress of science and technology, the increase in the means of action available to governments and the gathering speed with which they can apply them within the area of an ever more finite world, have not deprived international relations of their twofold aspect of competition and cooperation, but quite the reverse. The example of the International Bureau of Weights and Measures, the first international laboratory created by agreement between governments, throws some light on the nature of the changes which were to come about on the scene of international cooperation. The general adoption of the metric system induced countries to coordinate their systems of measurement, and to check and compare their national standards: the needs of science bring about the intervention of the authorities and the consequences of science institutionalise it. *On the international plane, as on the national plane, technonature is the situation created by the contingent requirements of scientists which must be brought under state control because of the political consequences of their work.*

The International Metric Commission, from which the International Weights and Measures Bureau originated in 1875, was the first occasion on which diplomats and politicians were associated with scientists. Associated, but no more: their role was still very discreet, since it was limited to facilitating, at government level, the establishment of institutional links between specialists from different countries. In sum, the diplomats were content to lend their services, and if they intervened it was in the final act of the negotiations and not in the negotiations themselves. It is easy to see why: in the nineteenth century science did not yet involve any major interest of states, and the organisations which were the subject of intergovernmental agreement had such limited budgets and objectives that they raised no great political problem. What brought scientists and diplomats together before the beginning of the twentieth century was above all the desire to improve channels of information within scientific circles themselves or the need to follow up consequences, from the point of view of international law, of certain discoveries and applications (such as quarantine and vaccination regulations).[17]

And yet the cooperation which the scientists professed to be seeking around the conference table was already coming up against the rivalries and competitions which are the daily bread of diplomats; the finite world is not yet the unified world—even with the help of the metric system. Since states intervene as such in international scientific relations, cooperation in this field must in future be looked at from two angles, scientific interest and political expediency. The two points of view may intersect, associate or merge, but they may also clash, as witness, for example, the history of the United Nations World Health Organisation—a domain in which political and scientific objectives might well be expected to harmonise without difficulty.[18]

After the first world war, private scientific associations, even if they raised no major political problem for governments, could not escape clashes with them, sometimes to the point of finding their purely scientific activity paralysed by political considerations. This was the fate of the two non-governmental organisations which were the forerunners of the International Council of Scientific Unions (ICSU); the International Association of Academies could not do its job because of the predominance of German institutions, or the International Research Council because of their exclusion.[19] After 1945 there were even more instances of the violation of the ideal frontier which protects scientific associations from political interference at international level; the line of demarcation between the zone of technical interests and the

17. See Jean-Jacques Salomon, *International Scientific Organisations*, OECD, Paris (1965), introduction, and 'International Scientific Policy', *Minerva*, 2 (summer 1964), 411–34.

18. See *The First Ten Years of the World Health Organisation*, WHO, Geneva (1958), 21–31.

19. See H. Spencer Jones, 'The Early History of ICSU, 1919–1946', *ICSU Review*, Rome, 3 (October 1960).

zone of political interests is becoming more and more an imaginary line.

The International Geophysical Year, for example, organised by the ICSU, which in 1957–8 coordinated the observations made by some sixty countries, was not immune from the pitfalls of national susceptibilities or from the vicissitudes of the cold war. When it was desired to set up 'World Data Centres' to store and disseminate the observations, it soon became apparent that only the United States and the USSR had the necessary financial and material resources to organise a centre and to process the data properly. The selection of these two countries, however justified on the grounds of economy, was in danger of consolidating the monopoly of power of the 'Big Two' over the international scientific community. On the model of any other international assembly where the spirit of cooperation has to be minted into current coin through the medium of judicious compromise, it was decided to set up a third World Centre, made up of mini-centres in various countries which would compile data in certain fields of science only. But no compromise could be found for the participation of the two Chinas. Communist China, which took part in all the preparatory work for the International Geophysical Year, had uttered the warning that it would withdraw if Formosa were to come in. For four years Formosa displayed no interest; finally, under American pressure, only two months before observations were due to start, its official request for participation was presented and accepted, justified by a small, hastily improvised scientific programme, and communist China immediately withdrew.

If the system of international scientific relations still has its two aspects, private and governmental, it is all the harder to distinguish them, since each of then now depends for its own fortunes, its resources, its interests and its objectives, on the other. Fellowships and government grants awarded to scientists to facilitate international exchanges are not disinterested, and what has been called 'scientific tourism' is not always dissociated from political or military espionage. Just as the missionaries backed up the settlers in conquering Asia or, more recently, the archaeologists seconded the chancelleries in carving up the Middle East, scientists are called up, officially or unofficially, to discharge the public functions which are favoured by their technical or private concerns. They have a part to play, too, in the big international fairs where the conquests of science are vaunted as national achievements. In the context of the new 'cultural diplomacy' which has been defined as 'the manipulation of cultural materials and personnel for propaganda purposes', scientists are trotted out like film stars or champion boxers.[20]

The first manifestation of the change in the nature of cooperation is to be found in its functions, in which governments and no longer scientific circles alone are now directly interested. The progress made during and after the war demands closer cooperation among scientists, making use of new institutional mechanisms, if only to make up the leeway caused by war damage and enemy occupation, especially in Europe. But it is no longer only a question of making up the leeway; science has changed its dimensions, and so has the power which it confers. Science now has far more branches, which is enough in itself to explain the great demand for men, equipment and money and the need to pool efforts. Each branch of 'Big Science'—atoms, space, computers—calls on its own for investments which are out of all proportion with those needed by research at the turn of the century. And yet, this is not the decisive fact: the 'new world' revealed by the Hiroshima blast is a political bipole, perpetuated by the capacity of two powers—and two only, pending the emergence of China—to develop all the most advanced technologies without having to rely on cooperation.

Europe, the cradle of scientific cooperation and once the core of the international system, is at the same time discovering that it is now only on the periphery of that system and that it no longer includes any country individually capable of undertaking on the same scale the research and development programmes pursued by the higher bipolar powers. Except in the spheres which by their transnational character call for pooled research (meteorology, oceanography), the 'Big Two' need not look to scientific cooperation for anything beyond what it has always yielded, namely the exchange of ideas and information. All other countries, in contrast, are condemned to look upon it as a roundabout but essential route towards achieving some of their national objectives faster or more cheaply. It follows from this that the conditions governing cooperative scientific action, and particularly the creation of specialised international organisations, are vastly different from what they were only fifty years ago; all these joint undertakings will be the subject of intergovernmental agreements, they will originate and develop in a political context and they will be obedient to political motives as well as to scientific considerations.

The dimensions of scientific cooperation have changed: it relates no

20. F. C. Barghoorn, *The Soviet Cultural Offensive*, Princeton, N.J. (1960), 10.

ionger to exchanges and communications addressed solely to specialists, but to the joint conduct of vast undertakings (high-energy physics, the industrial exploitation of nuclear energy, space exploration, and the like) within international organisations whose budgets vie with national research budgets. Its responsibilities have changed, too: whereas it originated and developed essentially under the impetus of scientific circles, it is now largely fostered and financed by governments through institutions whose objectives and activities are not exclusively scientific. Finally, it has changed direction: the costs and benefits of joint action are measured no longer by the general interests of science, but by the objectives set by each country under its own programmes of scientific, technological, economic or military expansion. A country's participation is thus regarded as one national investment among others or, in other words, as the instrument of a policy designed to complete or reinforce its scientific potential or in some other way to further wider aims.

The commonest characteristic of these cooperative research undertakings, however different the participating countries, the nature of their activities or their procedure, is also the one which best reveals their dual scientific and technical character: their supreme body, which has the last word on budgets and therefore on the development of research programmes, always consists of duly authorised delegates, ministers, ambassadors or senior civil servants. In those organisations whose activities are not purely scientific (Unesco, OECD, NATO, Euratom, etc.), heads of delegations are accompanied, or sometimes, in specialised committees, represented, by scientific experts. In organisasations with purely scientific or technological programmes (CERN, ESRO, ELDO), each country is represented on the council by two delegates, a diplomat and a scientist. There is even one international institution with a non-governmental status (EMBO) whose resources and main lines of action are decided by a ministerial conference which meets periodically for this sole purpose.[21]

The research requirements which compel international cooperative

21. On the political problems arising from and for these institutions, see Jean-Jacques Salomon, 'European Scientific Organisations', in *Science and Technology in Europe* (ed. Eric Moonman), Penguin Books, Harmondsworth (1968), 63–86. (Unesco, United Nations Education, Scientific and Cultural Organisation; OECD, Organisation for Economic Cooperation and Development; NATO, North Atlantic Treaty Organisation; Euratom, European Atomic Energy Community; CERN, European Organisation for Nuclear Research; ESRO, European Space Research Organisation; ELDO, European Launcher Development Organisation; EMBO, European Molecular Biology Organisation.)

action do not immunise scientists from national pressures but, conversely, the consequences of research, which limit or challenge the sovereignty of the nation-state, do not stop them from identifying with their own country. As E. B. Skolnikoff has emphasised, the specific property of certain modern technologies is that they have global effects which do not stop short at political frontiers; nuclear fall-out and satellite communications today, weather control and exploitation of the ocean depths tomorrow, the consequences of scientific enterprise have now attained planetary dimensions.[22] This does not mean that scientists of all countries form a common front to oppose those experiments which manifestly threaten to affect the environment or even, like 'Project Westford,' to compromise the possibilities of further research.[23] Quite the contrary: when faced with the demands of national security or the general policy options of their governments, they react no differently from their non-scientific countrymen; the universality of science creates no bonds which weigh the scales specifically in favour of identification with a transnational community. Unity of interests, inspiration and methods no doubt defines an intellectual community, but does not make it into a *res publica*.

The symbol of the human community

And yet it is because they enlarge the reality of communication into the ideology of cooperation that people expect science to be a mediating element in international conflicts. The good understanding which prevails at scientific meetings already indicates that technical consensus may lead to comprehension of one's neighbour, if not to his political conversion. The common language which is a vehicle of communication reserved to specialists is at the same time the pledge of an amity and confidence which should be capable of spreading beyond the walls of technical discussions. If history is what it is, with its heavy burden of ulterior motives, lack of understanding and conflicts, the reason is that the voices through which it speaks are those of unreason or lack of reason; since it is the fruit of defective communication, to influence

22. Eugene B. Skolnikoff, *Science, Technology and American Foreign Policy*, MIT Press, Cambridge, Mass. (1967), especially chapter 7.

23. Designed to check theoretical principles, the experiment consisted of placing in polar orbit a belt of copper wires which would act as radio reflectors. Unlike the case of 'Project Starfish' where nuclear fall-out constituted a hazard to human life, the scientists who protested against 'Project Westford', organised by the military, were concerned solely with the needs of research, on the grounds that the copper belt was in danger of prejudicing future astronomical observations. (See Skolnikoff, *ibid.*, 84–6.)

the channels of communication would be sufficient to bring it to the wished-for conclusion of universal comprehension. States embark on joint actions from two motives, cooperation and competition, the first generally as a means and the second as an end. If scientific procedures are taken as the model and the medium, there is every reason to think that, with the data of the problem reversed, it should logically solve itself by the suppression of one of its terms; if cooperation becomes the end, then competition disappears.

There are in fact two spheres in which the spirit of cooperation has prevailed over competition between states. The first, it is true, involved neither their sovereignty nor their economic or strategic interests: following the International Geophysical Year, the Antarctic Treaty, 'freezing' all territorial claims for thirty-four years, excluding all military activity and guaranteeing free access to the region for all scientific expeditions, is an example of diplomatic and political action based on purely scientific motives. The second example, on the other hand, is one of scientific action based on diplomatic and political motives, namely, the treaty banning high-altitude nuclear tests. On the one hand, we have a subcontinent which (so far) gives no ground for competition between states; on the other, planetary space open to competition tempered by the common fear of atomic fall-out. Science has battered a breach in the wall of competition, but the wall has not thereby collapsed. Even the limits of the nuclear test-ban treaty conform to the most traditional aspects of cooperation between states; what the member states of the 'atom club' renounce involves no sacrifice of their sovereignty.

It is no paradox to find scientists who profess to be free from all ideology in the field of their own expertise assigning themselves an ideological role in the international sphere. The same hypothesis of the neutrality of scientific proceedings which should guarantee the rationality of political decision points to the society of scientists as the ideal model of the human community:

> The Pugwash Conferences [wrote their Secretary-General] have shown that it is possible to apply the scientific approach, which has proved so successful in science and technology, to problems which are only indirectly related to science.[24]

The operational aspect of scientific activities in their own sphere warrants its extension to other spheres; functionalism introduces into the realm of method the same promise of unification that cosmopolitanism hold out in the realm of personal relations.

24. J. Rotblat, *Pugwash—a History of the Conferences on Science and World Affairs*, Czech Academy of Science, Prague (1967), 7.

This attitude embodies all the illusions of the 'functional approach' applied to the integration of political units. According to Ernst B. Haas, there are four ways of achieving political unity;[25] each of them, almost without transposition, is found as a reference in the themes developed by scientists when they hold themselves out as a community which generates universal comprehension and pacification. Just as the integration of nations demands that common criteria of well-being should take the place of criteria of power, so the canons of scientific activities should take the place of the disorder of national passions and rivalries:

> Science is today one of the few common languages of mankind; it can provide a basis for understanding and communication of ideas between people that is independent of political boundaries and ideologies.[26]

Just as experience acquired in a sector where the nations have succeeded in pooling their efforts provides a lesson for other sectors, so scientific cooperation should afford an example for other spheres:

> The creators of CERN will certainly regard it as a great reward if such cooperation not only leads to scientific progress, but also to a better understanding among the peoples of the world.[27]

Just as the task of integration is a job for technical experts rather than politicians, so the scientists, experts *par excellence,* are the best prepared to point the way to unity:

> The ability of scientists all over the world to understand one another and to work together, is an excellent instrument for bridging the gap between nations and for uniting them around common aims.[28]

Finally, just as the existence of a supranational authority should be the source of loyalties which transcend traditional allegiances, so science offers the best terrain for a 'holy league' of mankind which would overcome all national divisions:

25. Ernst B. Hass, *Beyond the Nation-State,* Stanford University Press, California (1965), 21–2, 47–50.
26. George B. Kistiakowski, 'Science and Foreign Affairs', *Bulletin of the Atomic Scientists,* **16** (April 1960), 115. There is such a wealth of quotations on these topics that they could almost be taken at random.
27. C. J. Bakker, 'CERN as an Institute for International Cooperation', *Bulletin of the Atomic Scientists,* **16**, no. 2 (February 1960), 57. CERN, of which C. J. Bakker was the first Director-General, is located at Geneva-Meyrin and is designed to study high energy particles.
28. Declaration of Vienna, Third Pugwash Conference, in Rotblat, *op. cit.,* 93.

It has often been suggested that, should the earth be attacked by Martians, the nations of the world would submerge their ideological difficulties and join to stave off the invader. In considering international cooperation in science, and its future development, one wonders whether the assault on science could not take the place of the repulse of the Martians as a common understanding which could draw nations together.[29]

In the extreme case, the scientific community affords the pattern, if not the beginnings, of the world government which alone can sweep away the divisions between nations, just as only the scientific conception of truth can unify man's minds. This is the theme of Einstein's letter to Freud before the second world war:

> Do you not believe that a free encounter of personalities, whose action and previous creations afford a guarantee of their abilities and of the purity of their intentions, could bring a remedy?[30]

It was also the theme of Father Dubarle who, while ruling out the possibility that the scientific priesthood might ever assume power, envisaged the possibility of a 'world authority' of which they would be *ex officio* members:

> Science is therefore the first world-wide human power which has emerged among mankind. At the present moment this power seems already *set up* for the universality of the earth and the human community. But the theologian would readily add that science does not yet seem to him to be a *confirmed* power in its universally human function.[31]

The political scientist would be more inclined to say that 'a method is not a policy, a process is not a direction'.[32] The functional approach no doubt demonstrates the advantages of what should be, compared with the disadvantages of what is, but it has its own disadvantage of leaving out of account what we must call, with Stanley Hoffman, 'the tyranny of reality': the nation-state is still with us and the New Jerusalem is still to be sought. The canons of science as method, the skills of scientists as experts and the bonds of the scientific community as model are not enough to convince the human race of the logic of unification, any more than the evident functionalism of integration has been enough to

29. Alvin W. Weinberg, 'Progress in International Science', *Bulletin of the Atomic Scientists*, 14, no. 10 (December 1958), 404.

30. Albert Einstein, translated from *Comment je vois le Monde*, Flammarion, Paris (1934), 64.

31. See D. Dubarle, 'Towards a World Community of Scientists', *Bulletin of the Atomic Scientists*, 15, no. 5 (May 1959), 179–80, and the *Minerva* article already cited, 425. The quotation is taken from his *La Civilisation et l'Atome*, Edition du Cerf, Paris (1962), 241 (italics in the original).

32. Stanley Hoffman, 'Obstinate or Obsolete? The Fate of the Nation-State and the Case of Western Europe', *Daedalus*, 95, (summer 1966), 886.

outweigh national diversities and rivalries. The cosmopolitanism of scientists stems from links which are fostered by the structure and needs of scientific research and not from any specific power of science to generate solidarity among men in the same way that it can obtain their consensus.

The internationalist spirit displayed by scientists as individuals does not automatically make them any more cohesive as a body, and the forces which unite them technically are no guarantee of their political unity. The international character of the intergovernmental institutions active in scientific research in Europe is the result not so much of a specifically international, or even European, attitude on the part of scientists, as of the absence of national scientific institutions capable of carrying out the same activities on the same scale. But there are far more scientists in Europe who would choose to escape from the shortcomings of their own country by integrating with a new national community (preferably the United States) than by responding to the call of internationalism.[33]

Prometheus bound

Faced by the failure of the League of Nations and the threat of world war, Einstein visualised recourse to the intellectual community as a means 'of exerting an important and morally salutary influence on the solution of political questions'.[34] The idea of the social and political responsibility of the scientist did not yet imply the idea of his responsibility as researcher. Whatever changes the progress of science might make in the course of history, they were not calculated in any way to call into question the course of science; the scientist was duty bound to enlighten the authorities, as one intellectual among others whose special skill or wisdom justified intervention in the political sphere. The aftermath of the second world war redefined the situation of the scientist in his relations with power; he emerged as the source of the major political problem facing modern societies—that of their survival.

The ideology of science, looked at solely from the point of view of the truth which it pursues, cannot withstand the power of destruction and manipulation which research confers by virtue of its growing associa-

33. See Albert H. Teich, *International Politics and International Science: A Study of Scientists' Attitudes*, thesis, Department of Political Science, MIT, Cambridge, Mass. (May 1969), especially 365–6.
34. Einstein, letter to Freud, translated from *op. cit.*, 65, note 30.

tion with the state. Only the merest technician of knowledge could profess to disregard the reasons why the whole function of the scientist
invests him with special responsibilities not shared by other intellectuals.

> We men of science, whose tragic fate has been to help in the creation of
> more horrifying and more efficient methods of annihilation, must consider
> it our solemn and supreme duty to do all in our power to prevent these
> arms from being used in attaining the inhuman goal for which they were
> invented.[35]

For Einstein, the concern to exert an influence as one intellectual among
others has become an obligation which the scientist as such can no
longer evade. The consequences of the scientific enterprise, the function
fulfilled by researchers and the power they hold in the political area of
technonature have swept away the ideal frontier traced by the ideology
of science between the practice of research and its social implications.

But what exactly can scientists do to imprint *in concert* the mark of
the rationality of science on the irrationality of history? Common
language does not necessarily mean common understanding.

> We must revolutionise our thinking, revolutionise our deeds and have the
> courage to revolutionise the relations between the nations of the world.[36]

Neither does it change the old man into a young one; science changes
the world, but it does not change men, and history takes its course without the efforts of the scientists to modify it in the political sphere having
any greater or different effect from those of non-scientists. Appeals to
public opinion or 'international conscience', approaches to governments
or to the United Nations, conferences and seminars of specialists ending
in petitions or recommendations, all these initiatives lie in a territory
where scientific endeavour has no greater power to establish consensus
than non-scientific endeavour.

The Pugwash conferences, assembling scientists 'of good will' in a
private arena, did not escape the restraints of technonature, the benevolence or complaisance of governments towards them being taken for
granted.[37] The political function of scientists is continued in these

---

35. Einstein, translated from *op. cit.*, 183, note 3.　　　36. *Ibid.*, 185.
37. See D. Dubarle: 'It is quite clear, for example, that Russian participation is the subject of consent which goes beyond individual initiative. What is
true of Russian participation is no doubt equally true, in a different form, of
American participation.' (*La Civilisation et l'Atome, op. cit.*, 40). It is, indeed,
enough to read the list of those who have attended the different Pugwash
conferences to recognise the names of scientists and political scientists connected officially or unofficially with the political decision-making bodies in the
west as well as in the east.

meetings, for all that they are professedly intended to purge the scientific approach of everything it shares with political approaches; the fact that the talks are private and remote from the corridors of power does not prevent some of the participants from 'living and symbolising' international relations like diplomats or soldiers. And all of them, whatever their wishes or their acts, are faced with the contradiction that, as citizens like any other, they embody the political entity to which they belong and, at the same time, as representatives of science, the intellectual entity to which they profess allegiance.

Torn between their obvious special responsibilities and their inability to discharge them by specific means, they can at least exercise a magistracy of influence—'enlightening princes, educating peoples'.[38] Some of them try to convince governments by personal intervention (like Niels Bohr sending a letter to the United Nations); others try to gain international support by addressing themselves collectively to public opinion or governments (for example, the petition by Linus Pauling presented to the Secretary-General of UNO with the signatures of 9,235 scientists from forty-four countries, including thirty-six Nobel prizewinners). The duty of information springs both from skill and from responsibility; it is a recognition of the political character of the scientific enterprise in so far as its ends are oriented in relation to the complex of political decisions which treat it as a means. But this dialogue between skill and world opinion, like the dialogue with authority at national level, immediately comes up against the limitation that it assumes a neutrality which the whole situation of scientists in the social system renders suspect; you cannot be 'above the fray' when what you do, what you discover and what you invent determine the instruments of the fray and the technical conditions of its preparation and its course.

Prometheus was bound, says the legend, because the gods were jealous. In the eyes of science, the gods are dead and the human tribunal has descended from heaven to earth and has become secularised, like science. The Prometheus of the modern world, the scientist binds himself in his own chains by the mere fact that, whether he likes it or not, his vocation hurls him into the political arena; the pure sky of theory, indifferent to the contingencies of history, belongs to an age of science which has been swept away by its own operational efficiency. On the ideological plane of objectives, the scientist cannot claim any greater

38. J. Meynaud and B. Schroeder, *Les Savants et la Vie internationale*, Etudes de Science politique, Lausanne (1962), 82.

competence than the man in the street, whatever privileged access he may have to the technical data which should inform opinion or brief the decision, and, *at the same time*, he cannot shelter behind the neutrality of science as a pretext for shutting his eyes to the functions he fulfils in technonature. The international scene of science displays the same contradictions as the national scene; the only difference is that, since it is played out on the theatre of the world, the profession of neutrality is all the more unable to keep its promise of universality when it is applied to anything which is not science itself.

Science has the initiative in the problems it raises for the modern world, but loses the initiative when it comes to solving them. The movement from means to ends involves presuppositions and choices in which neither the scienticity of science nor even the spiritual community of which scientists boast is of any help; the monopoly of knowledge, if it goes beyond the field of technical applications, becomes part of the common lot of individual commitment. The scientist does not feel himself equipped to live either 'for' or 'by' politics, and yet he is condemned to live 'in' politics; the modern Prometheus is bound by his vocation to the ambiguities of a role which neither his training nor his profession have prepared him to play.

Those scientists who still see science as something more and better than a technique gain no wider choice of alternatives by appealing to the whole world than by trying to influence the national decision-makers; either science must be assigned the mission of throwing light upon and even of formulating the ends to be pursued by power, or power must be charged with defining the ends to be pursued by science. Illusion or desertion, in either case the scientists remain chained—bound—to the decision-making system which determines the resources that will enable them to fulfil their vocation. Protesters but part of the establishment, or established even though protesting, they are in any event trapped in the pitfalls of political life. In the extreme case, the outcome can only be either to turn a deaf ear to the tumult of the world and devote oneself wholeheartedly to research, or to sacrifice one's vocation on the altar of militant action.[39]

39. This is well illustrated by the statements of three Harvard biologists who succeeded for the first time in isolating a pure gene; in announcing their success, they seized the opportunity of their press conference to utter a warning against the threat of 'genetic engineering'. In a letter published in reply to the protests aroused by this 'political' intervention in a scientific exposition, they had no hesitation in writing: 'If our arguments mean that the progress of science itself may be interrupted, that is an unfortunate consequence we will have

Modern societies are not founded on scientists any more than ancient societies were founded on philosophers; it is the technicians who determine the decisions, and the magistracy of influence never has more than a very limited political effect. But it is a consolation that there are always scientists who refuse to cut down their conception of science to a purely technical adventure and will not allow it to be merely one instrument among others in the service of politics. The mystifying element in the ideology of science does not prevent it from leavening the behaviour of individuals who cling to the ideal of an intellectual community determined to question the meaning of the ends served by the creations of the mind. One may smile at the pacifist utopias which are the pabulum of the Pugwash Conferences or at the political naivety displayed by many of the articles in the *Bulletin of the Atomic Scientists*, but the fact remains that those who lift their voices there have not betrayed their vocation by blindly accepting the political implications now inherent in their vocation itself: 'They saw it their duty to try anyhow . . .'.[40] People may perhaps forget tomorrow how much anguish it cost some of them to discover that the pursuit of knowledge is not an innocent adventure. However vain their efforts may have been to bring the ends of power into closer harmony with the ends of knowledge, they at least have saved the honour of science.

---

to accept. It certainly should not inhibit us from speaking out on crucial issues.' (*Nature*, 224 (December 1969), 1337.) One of them, Dr Jim Shapiro, decided 'to quit science for politics', especially because he 'believes that the work he does will be put to evil uses by the men who control science—in government and large corporations—in the way that atomic energy, for example, was put to evil uses'. (See *Science* (13 February 1970), 963.)

40. E. Rabinowitch, 'The Bulletin of the Atomic Scientists', *Impact*, 3, no. 2, Unesco, Paris (summer 1952), 103.

Conclusion

Science has been furthered during recent centuries, partly because it was hoped that God's goodness and wisdom would be best understood therewith and thereby—the principal motive in the soul of great Englishmen, like Newton; partly because the absolute utility of knowledge was believed in, and especially the most intimate connection of morality, knowledge and happiness—the principal motive in the soul of great Frenchmen, like Voltaire; and partly because it was thought that in science there was something unselfish, harmless, self-sufficing, lovable and truly innocent to be had, in which the evil human impulses did not at all participate—the principal motive in the soul of Spinoza, who felt himself divine, as a knowing being: —it is consequently owing to three errors that science has been furthered!

NIETZSCHE

The honeymoon days of the alliance between knowledge and power, which have lasted since the second world war, are drawing to a close; nearly everywhere, the rate of public research expenditure is beginning to slow down and budgets to level off and even to shrink. This expenditure is nevertheless still substantial, both in absolute figures and as a proportion of gross national product, and there is no reason to think that industrial societies are getting ready to cut down on research. But the years of blind affluence are over; governments have learned to be more selective in allocating resources to science or, in other words, more careful to invest in those research activities whose results they deem likelier to have a direct influence on the attainment of their short-term goals. The mystique of the last quarter of a century amounted in sum, to proclaiming that what is good for science is necessarily good for society, governments being all the more willing to endorse this profession of faith by the scientists in that they regarded science as an inexhaustible cornucopia from which, as an ineluctable effect of the growth rate of their total expenditure, there flowed an endless stream of instruments of power, prestige, economic growth and well-being.

Whatever the disequilibrium between disciplines or research sectors, science did very well out of this period, if only by the growth in the number of researchers and in research facilities. De Solla Price was not wrong, however partial and approximate the data for his calculation, in writing that the growth of science 'is something very much more active, much vaster in its problems than any other sort of growth happening in the world today'. Whether measured by research expenditure over a quarter of a century or by the number of specialised reviews, which have risen in fifty years from some ten thousand to more than a hundred thousand, 'the density of science in our culture is quadrupling during each generation'.[1] Recognised as a 'national asset', priority investment, 'imperative for a modern state', supported as a productive force whose promises are fulfilled in the short term, elevated to the rank

1. Derek J. de Solla Price, *Science Since Babylon*, Yale University Press (1962), 107–8.

of an economic activity whose budgets and results stimulate the whole of the economy, it has lost its status as an isolated institution in the midst of our society, an intellectual or cultural enclave reserved for an élite and without any direct influence on world affairs, to become an industrial enterprise inconceivable without vast resources of organisation and concentration, exactly like any other modern production activity.

In becoming organised labour, scientific research yields willy-nilly to the imperatives of organisation which determine the forms and pace of modern production; there is now little but the uncertainty of its time-lags and its results to distinguish it from other forms of production. The distinction is nevertheless an essential one which stops us from taking literally the insistence on economic return which appears so often in official dicta on science policy and even in essays on the subject by scientists who hold, or who have once held, political responsibilities: research is always *research*, that is to say, an activity which is always something of a gamble owing to the uncertainty of its results, no matter how high the stakes may be raised and however finely they may be calculated.

In the mind of governments, which regard it as a miraculous treasure chest, as well as of the scientists who offer it as such, science is closely associated with the dedication of modern societies to continous growth, when it is not deemed to be the direct cause of that growth. But this mystique gives birth to myths, and the policy which inspires it gives rise to consequences which cast some suspicion on the reality, or at least on the inevitability, of the resultant progress. The positivism which inspired the institutionalisation of the relations between science and the state has its limits, no less than other forms of positivism—starting with those of a naïve faith in the benefits which both partners should derive from their irreversible alliance. The alliance has not lacked a counterpart: the state has created its clientele among scientists, but it depends ever more closely on the results of scientific research; this clientele, for its part, is not wedded to the goals of power beyond the point at which it ceases to profit from them in attaining its own goals. In discovering at least sufficiency, if not opulence, scientists could not imagine that the goodwill of power towards them was grounded on utilitarian requirements which they repudiated (or professed to repudiate) in the name of an idea of science which the age of organisation and industrialisation has banished to nostalgia for the past.

Another positivist myth was the idea that the results of research activities can be measured like the investments on which they depend.

The economists who, a few years ago, thought to render mathematical account for the output of research are today asking how far the research system really contributes to economic growth and are discovering that the process of technological transfer is anything but clear, and in any event very remote from the model of cause and effect which some of them hoped to construct in the 1950s. The countries which devote the largest proportion of their gross national product to research activities are not necessarily those with the highest growth rate, and science is no pledge of economic development merely because it absorbs a 'magical' proportion of national resources. The gap between research activities and other production activities may well be narrowed, but it is none the less impossible to bridge unless the process of discovery can be purged of every random element and innovation of its dependence on the capacity of a social system to exploit research results rather than on the talent of researchers and the sum of their activities. Recognised and established as an economic subsystem of the social system as a whole, the scientific research system is not thereby reduced to a production activity like the rest.

Finally, and above all, there is a myth of the most summary positivism, the idea that scientific affairs, because they affect science, must display greater rationality than the other fields in which the state wields authority. Programming methods have progressed and will go on progressing more and more, but the planning of research, in the sense of a set of provisions consistent in all rationality with the measured needs of society and following a balanced allocation of resources, remains a pious hope—or a piece of humbug—of administrators enamoured of technocracy or scientists enamoured of quantification. Choices affecting research activities cannot be more rational than any others, whatever mathematical instruments are used, because there is no possible way of measuring the rationality of decisions; if one sector gets more support than another, it will never be for objective reasons which the bureaucrats may claim to have worked out, but because the needs and possibilities are and always will be, in the last analysis, evaluated at best as calculated risks and at worst as the result of a compromise between conflicting alternatives. If rational decision should be the aim, it is never attained in practice, unless we live in a Hegelian history where reason always justifies its protagonists after the event.

Born of the last world war and of its far from peaceful aftermath, science policies everywhere confirm the recognition of the instrumental character of knowledge as a collective investment whose primary justifi-

cation has nothing to do with its intellectual purpose. It is science fulfilled as a technique which interests the state and justifies in the eyes
of public opinion the money spent on research, and not science as culture aimed at the enlargement of knowledge for its own sake. The
adventure of scientific research has yielded more than Bacon hoped
when he foreshadowed the rise of knowledge fulfilling itself as power; it
has culminated in the accumulation of material goods which he pre-
dicted as the harvest of the accumulation of new knowledge, but only
at the cost of a subordination of new knowledge to material goods
which his 'programme', far from regarding as foredoomed, denounced
in advance:

> Truth and utility here are absolutely the same thing; and the works them-
> selves have greater value as pledges of truth than as contributions to the
> comforts of life.[2]

The support which society accords to scientific knowledge looks not so
much to the 'experience of Light' as to the 'experience of Fruits'. Hence
arises that misunderstanding irreversibly inherent in the whole nature
of the relations between knowledge and power: scientists pursue (or
profess to pursue) the light, thanks to the support of society, while
society, and especially the state, supports the scientists mainly with an
eye to the fruits of their work. Science as an ideological concept, vested
with its own ends and its own values, is diluted in the functions it discharges as one link in the chain of the industrial system organised for
the exploitation of knowledge.

And no doubt, whatever line is taken to justify the money spent on
it, the result is bound to be the same; science is no longer a heterodox
social institution in the economic system of production and consump-
tion, in the sense that there are and always will be forms of creation
whose product, both in intention and in use, is irrelevant to the process
of industrialisation. Quite the reverse, it affects this process as a decisive
force in the metamorphosis of modern societies into 'scientific societies',
if only by the constantly more specialised technical qualifications demanded by the management of these societies deeply coloured by the
general adoption of scientific methods and paraphernalia. The young
man who by frequenting museums or reading poetry discovers his vocation as a painter or a writer can always do without state support in
fulfilling that vocation; the scientific researcher cannot manage without

2. Francis Bacon, *Novum Organum* (ed. Spedding-Ellis), book 1, aphorism
cxxiv, reproduced in *Selected Writings* (ed. Hugh G. Dick), The Modern
Library, New York (1955), 534.

state support. The intellectual adventure of science is no longer an individual odyssey but a joint venture whose artificers, interests and directions largely depend on the choice of the body social as a whole.

Politics enters into every aspect of science which is not pure theory: in this sense, 'pure' mathematics, in its relation to history, is always nearer to art than to production. But, in another sense, even 'pure' mathematics takes its stand, like all the experimental sciences, in the area of technonature: the effort of scientific education needed by the modern research system entails political measures of support and promotion at every link in the chain of the process of discovery, even those which seem to hold out least promise of leading to practical applications. The country which wants a massive 'output' of good mathematicians must know how much it is prepared to pay to train them at all levels of mathematical research, from the most abstract to the most immediately applicable.[3]

Dependent on the goals of power, science at the same time determines the range of options open to it; if its claim to autonomy applies and will always apply to the slightest threat of authority over the content of its activities, science regarded as a social institution is by no means the apolitical sanctuary in which scientists believe or wish or claim that their profession can shelter. Technonature is the inevitable meeting ground between the ideology and the scienticity of science which misunderstandings, mental reservations and professions of neutrality do not prevent from often amounting to complicity.

II

If science fulfilled as a technique is one means among others of attaining political ends, the problem it raises is no longer that of the ends

3. As demonstrated by a report of the National Academy of Sciences which emphasises how greatly the rise of mathematics in the United States depended on support from the Federal Government: 'Before World War II, the United States was a consumer of mathematics and mathematical talent. Now the United States is universally recognised as the leading producer of these.' This song of triumph, a fine anthology piece of nationalist science, is supported by very convincing statistics on the growing contribution of the United States to international mathematical conferences, the number of Fields medals won since 1945 by American citizens or 'long-time residents of the United States [who] should now be considered to be members of the US mathematical community', and on the growing proportion of references to American journals in specialised European journals such as *Acta Mathematica*, *Commentarii Mathematici Helvetici* and *Mathematische Annalen*. (*The Mathematical Sciences: a report*, Committee on Science and Public Affairs, National Academy of Sciences, Washington (1968), summary, 11–12.)

which it sets itself but that of the ends which it serves. In other words, can science policies have priorities any different from those they have had so far, entailing by the same token a different allocation of the majority of scientists? And how far can such a question be answered, even if only in part, from the scientific side and not from the political side alone? It may be downright absurd, because it is quite unrealistic, to stage the debate on this ground; the order of priorities displayed by science policies stems from what is fundamentally unreasonable in human history rather than from what is irrational. It can always be said that if our societies choose to develop and use science to increase their power and glory rather than to solve their internal difficulties, improve the living conditions of their privileged citizens and help to feed the underprivileged two-thirds of humanity, it is because it is human nature to exploit the ways of reason for the profit of unreason.

At the end of his excellent study of the role of invention in economic growth, Schmookler admits his resignation to the growing subordination of research to this will of society:

> If modern men at the same time see technological progress as an inevit-
> able consequence of the endowment of life with creative intelligence, it is
> only because they forget the examples of the ascetics, on the one hand,
> and of those medieval craft guilds that for centuries suppressed invention,
> on the other. From this vantage point, the pervasiveness of technological
> progress in modern times appears not so much a consequence of man's
> nature, though it certainly expresses an important aspect of it, but perhaps
> even more an outgrowth of the importance attached to material wants in
> modern times and of the indefensible military position of societies centred
> on other values. The increase in our capacity to wage war, thermonuclear
> and conventional, which has accompanied the increase in our capacity to
> satisfy our private wants may in the end justify the position of the
> ascetics.[4]

Even so, we may ask how far this will of society is, by virtue of scientific and technical progress, an adventure chosen rather than merely suffered. The idea of a science policy suggests that the decision-makers at least intend to orient the research system towards the attainment of certain goals, but are these goals themselves not in fact defined by the momentum which the system itself has gathered? What science deems feasible—'the technically sweet'—becomes in technonature something which power cannot do without, not because it meets a real need but because it determines the artificial needs of society. Because it is feasible, because it can be done, it must be done, even if the result creates more problems than it solves or even if it solves nothing at all.

4. J. Schmookler, *Invention and Economic Growth*, Harvard University Press (1966), 214.

This derisory state of affairs is nowhere more evident than in the application of science and technology to the strategic questions of the nuclear age. Since each side had enough force to act as a deterrent, but never enough to stop the other side from hoping to gain the initial advantage, the hope of influencing the 'balance of fear' by increasing the number of nuclear missiles or the destructive force of the warhead is doomed to perpetual disappointment. Whether there are a thousand or a hundred thousand bombs, whether they are counted in hundreds or thousands of megatons, whether the guided systems are more numerous or faster, there will always be enough to enable the side which is first attacked to hit back. And it is this margin compared to the total number of atom bombs in reserve—so out of proportion to the aim pursued that one can speak of overkill—which determines the real balance, each side always having enough to deter the other, whatever progress may be made in perfecting new systems of attack or detection.[5]

Nowadays, the race in antiballistic systems leads to the same kind of gullibility; even if the technical problems were solved (which, in the case of the *Safeguard* system, and even its minor version, *Sentinel*, is far from certain), it would be enough for the Soviet side to recognise that the Americans were trying to achieve what they regarded as feasible for the Russians in turn to conceive other systems of attack or defence as feasible and therefore set to work on them.[6] This chess game is bound to end in stalemate; technological development swallows up the whole content of strategic concepts for lack of political imagination to substitute other goals. But the prospect of a stalemate—a drawn game, wiping out all gains and losses—does not do away with the threat, with the result that where this is set as the sole objective there is no way of stopping the escalation; the game is only apparently drawn, and the threat survives in spite of the gigantic resources invested in the effort to diminish it. In the words of Herbert F. York, one of the leading experts in this field:

> The arms race is not so much a series of political provocations followed by hot emotional reactions as it is a series of technical challenges followed by cool, calculated responses in the form of ever more costly, more complex and more fully automatic devices.[7]

5. See Jerome B. Wiesner and Herbert F. York, 'National Security and the Nuclear Test Ban', *Scientific American*, 211, no. 4 (1964), 27–35.

6. On antiballistic systems, see *ABM: Evaluation of the Decision to Deploy an Antiballistic System* (ed. Abram Chayes and Jerome B. Wiesner), The New American Library, New York (1969).

7. Herbert F. York (former scientific chief of the ARPA—the Advanced Research Projects Agency of the Defense Department—and subsequently direc-

An endless process, apparently impossible to master and, above all, in
no way solving the problem it is reputed to solve:

> There is no technical solution to the dilemma of the steady decrease in our
> national security that has for more than twenty years accompanied the
> steady increase in our military power.[8]

Is it surprising that this confession of failure, witnessing 'the futility
of searching for technical solutions to what is essentially a political problem',[9] has not been transposed from the defence sector to the civilian
sector? It may be said that such a transposition is not legitimate because the issues are of a different kind, and above all because the game
is not played between two opponents or in a similar context; it is the fact
that they are doomed to parity which governs the rules of the strategic
game between the two 'Great Powers', but modern societies, when they
are industrialised and, even more when they belong to the Third World,
are strangers to parity of economic and social development. But the
rules of strategic competition have ended by imposing themselves even
in fields which have, apparently, nothing to do with the military field.
In practice, with many research projects it is impossible to distinguish
those which are concerned with defence from those which meet other
social demands.

In this context, the most urgent problems of mankind—overpopulation, underdevelopment, famine, health, education—are the poor relations of research activities; technical superfluity always takes priority
over human necessities.[10] If there is a relation between the armaments

---

tor of defence research and engineering in the same department, and twice a
member of the Science Advisory Committee to the President of the United
States before becoming Chancellor of the University of California at San
Diego), 'Military Technology and National Security', *Scientific American*, 221,
no. 2 (August 1969), 27.

8. *Ibid.*, 29.   9. *Ibid.*, 17: the formula is repeated on page 28.

10. Three regions (the United States, USSR, Europe) possess more than
80 per cent of world resources of scientific and technical personnel, although
they include less than 20 per cent of world population; less than 5 per cent of
total research and development activities is attributable to the underdeveloped
regions, which nevertheless include more than half the world's population; and
less than 2 per cent of research and development in the industrialised countries
directly concerns the problems of these regions. Neither the international concentration of research nor the nature of research carried on in the industrialised
countries operates to favour the development of the backward countries. Quite
the reverse: the orientation towards an economy founded on the exploitation of
technical innovations rather than of raw materials can only increase the speed
at which the gap widens (See C. Freeman and A. Young, *The Research and
Development Effort in Western Europe, North America and the Soviet Union*,
OECD, Paris (1965), 66.)

race, whose products and technological refinements seem to obey the
law of their own development by a sort of automatism, and the race for
technical innovation which is constantly being renewed within constantly receding limits, it is to be found in science policies in so far as
they aim at the accumulation of knowledge and technical progress without questioning the urgency of the problems they leave unanswered.
The end of 'laissez-faire' in the relations between science and the state
has not been the occasion for a reappraisal of the ends which science,
reduced to the rank of a technique, makes it possible to attain; it has
culminated rather, in the excellent phrase of an American left-winger,
in *'laissez-innover'*.[11]

There is no technical solution to essentially political problems in the
civilian sphere any more than in the military; scientific and technical
progress by itself solves no human problem, it merely shifts the solution
or raises fresh problems. The optimist regards its adverse consequences
as a series of 'externals', in the economic sense of the word, which the
inadvertence of the body social prevents it from foreseeing or forestall-
ing according to the methods of scientific enquiry;[12] the pessimist, in
contrast, will regard them as a form of destiny inherent in the accumulation of material goods which rules out any hope of mastering the process of technical change.[13] It is nevertheless the first time that the Sunny
Jims and the Moping Minnies of technical progress have been at one
in their concern at the adverse consequences and in looking for a human
meaning in what seems to be more and more devoid of one.

Modern societies have not been liberated by science, as the century
of the Enlightenment dreamed, in the sense that scientific research was
to pave the way for, if not to fashion, happiness, fulfilment and the
chance of a better order. They are much rather bewitched by the prestige of science and technology to such an extent that no vantage point
can any longer be found in the world which escapes the spell and from
which it can be denounced; a triumph or ruse of reason which ends
by giving the irrational the guise of reason itself. The age of science
triumphant is also the age of absolute menace, of nonsense, of derision;

11. John McDermott, 'Technology: The Opiate of the Intellectuals', *The
New York Review of Books*, 12, no. 2 (31 July 1969), 27.
12. See Emmanuel G. Mesthene, 'The Role of Technology in Society', 4th
Annual Report of the Program on Technology and Society of Harvard University (1967–8), 51 (reproduced in *Technological Change: Its Impact on Man
and Society*, Mentor Books, New York (1970), 40).
13. *Cf.* Jacques Ellul, *La Technique ou l'enjeu du siècle*, Armand Colin,
Paris (1954).

the conquest of rationality is crowned by the anarchic escalation of knowledge conceived solely in its instrumental function which it already seems beyond the power of knowledge to hold back or to direct.

The price of the laissez-faire which prevailed over the blossoming of the capitalist economy was the poverty and injustice of which part of mankind bore the brunt. The price of 'laissez-innover' threatens to be incomparably higher when we think of the global, planetary character of the new technologies and their consequences for man and his environment. From thalidomide to the Torrey Canyon, from nuclear fall-out to water pollution, the universal diffusion of technical progress has left no continent, no society, no class immune from effects both undesired and unforeseen, as though the spread of technology had the epidemic character of the great plagues and pestilences of the Middle Ages. As in the case of laissez-faire, it is true, the indictment of the societies which favour 'laissez-innover' exhibits only the evidence for the prosecution: insecticides and pesticides, for example, are nowadays cited only for the damage they do to ecological equilibrium and not for the human lives they have helped to save, and people are not very far from blaming Pasteur's conquests for the underdevelopment which is linked with overpopulation! But intent is not the only issue; the time is past when it was an article of faith to regard science as the means of 'freeing man from the restraints of nature' solely by virtue of its progress and its applications. These applications seem to have been more liable to enslave man in the bonds of technology.

The triumph of the artifices of anti-nature over natural phenomena merely shifts into the realm of man the sense of adventure which, in the days before modern science, men sought in the realm of the gods, but that adventure nevertheless still retains the aspect of an inexorable and absurd fate.[14]

> In the past [says Harvey Brooks, very rightly] we have permitted technology to proliferate in response to demand for its benefits on the assumption that technology was innocent of undesirable social, environmental or biological side-effects until proved guilty by clear evidence. Now there is increasing pressure towards shifting the burden of proof. The growth

14. An eloquent example of this absurdity is the atomic waste created by peaceful as well as military uses of nuclear energy: in the United States the capital outlay for the necessary disposal facilities runs into hundreds of millions of dollars and the maintenance of the 'sepulchres' costs over $6 million a year. As Lord Ritchie-Calder points out: 'There, in the twentieth century Giza, it has cost more, much more, to bury live atoms than it cost to entomb the sun-god Kings of Egypt'. (See Lord Ritchie-Calder, 'Mortgaging the Old Homestead', *Foreign Affairs*, 48, no. 2 (January 1970).)

or proliferation of technology is to be inhibited until it can be proved innocent of side-effects.[15]

Suspicion, says Brooks, should fall only on technology conceived as the utilisation of science, and not on science itself, even though it is the source of technical applications; each of them in fact feeds on the other, and the 'technically sweet' complex, which demands that anything which is feasible must be done at all costs, is just as compulsive for scientists as it is for engineers. The coincidence of interest between knowledge and power makes it impossible to 'whitewash' scientists from the responsibilities which are too readily laid on engineers alone: in technonature science activates and manipulates no less than technology.

In any event, this suspicion has given birth to the idea of 'technology assessment' which some scientists and politicians are more and more coming to regard as an essential function of science policies. It is visualised as a mechanism for 'mastering' the course of technical progress by forecasting and eliminating those of its consequences which are undesired:

> . . . to maximise our gains while minimising our losses. The challenge is to discipline technological progress in order to make the most of this vast new opportunity.

These are the words of the recent report on this subject by the National Academy of Sciences; the task is to place, or replace, the innovation process in the context of its social implications, and for this purpose to develop research programmes designed 'to minimise technological surprise and to deal more rationally with the burdens of uncertainty'.[16] A Herculean task: even if it were scientifically feasible, it still presumes that the political decision-makers would pay attention to it, at the cost of setting limits to the expansion of science and technology. How, for example, can institutions, committees and the like be set up which are so immune from vested interests and pressure groups that they can not only denounce the foreseeable or thinkable 'surprises' of technological development but also induce the authorities to take the necessary safeguards? It is no accident that the report, whose institutional proposals are somewhat unconvincing, speaks of a sort of 'Ombudsman' to lay

15. Harvey Brooks, 'Remarks on Technological Forecasting in the Next Decades', paper (stencilled) presented at a seminar of the Brooklyn Polytechnic Institute (autumn 1969), 12.

16. *Technology: Processes of Assessment and Choice*, report of the Committee on Science and Public Affairs of the National Academy of Sciences, Washington (June 1969), 12, and summary, 115.

down the law in the event of clashes between technical innovation and society.

It is precisely the expansion of scientific research which has made the race for technological innovation the battle-cry of international competition; it is hard to imagine the modern economy, based on profit and the satisfaction of artificial needs, sawing off the branch along which it is developing or slaking its thirst for innovation. But it is even harder to imagine the authorities, who play a decisive part in supporting research into new technologies, being able to withstand the pressure of international competition by setting limits to that research.[17] It is one thing to *react* against the 'adverse' consequences of the spread of a new technology, and quite another thing to *act* to forestall them by finding some sanction for technological development other than competition.

Technology assessment is a function which, in sum, recalls that of consumer protection associations which aim to identify those of the new products which best satisfy the buyer or deceive him the least. But the planning and spread of new technologies is not confined to consumer goods submitted to the choice of the consumer, deemed free and sovereign under the market economy; they are imposed on the community by decisions in which it shares, if it shares at all, only indirectly and remotely, and their consequences can be measured not in overall terms of profit and loss, but by the benefits derived by particular groups or interests in comparison with the cost or damage simultaneously incurred by others.

Furthermore, while science and technology must be summoned to

17. It is significant that the reservation of the industrial member of the group which drafted the report was so categorical that it is given in an annex: it emphasises that any mechanism designed to discipline technological development is in danger of 'irreparably damaging the systems of innovation that it is designed to stimulate and guide' (148–50). Even more significantly, the question of military technology is relegated to the end of the report, almost as an afterthought, although, as the authors stress, it accounts for more than half of the public resources assigned to research and development, and its results are almost automatically transferred to civil applications. Hence this marvellous statement: 'Our remarks concerning the proponents' control of crucial information for technology assessment are not intended to suggest that the proponents of military technology have used this control irresponsibly or to the detriment of the larger public interest, although a growing number of people believe this to be the case. We wish simply to point out that the present system for the assessment of military technology violates most of the canons suggested by the panel with regard to the representation of affected interests, the consideration of larger social and environmental contexts, the maintenance of future options and public visibility and review of the crucial information and arguments.' (113.)

make their contribution to the fight for 'a better environment' or 'quality of life', this objective is in any event less easy to identify and the means of attaining it are less tangible than if the goal set is a given technological programme. It cannot in any event mobilise scientists, or even attract their efforts, in the same way as major research programmes stimulated by international competition. In the absence of this stimulus, research activities would immediately cease to enjoy the favours of the state with the same largesse. It is true to say that no one knows whether a moratorium on research designed to multiply technical innovation would lead to economic stagnation or recession, but 'the tyranny of reality' or, in other words, the inevitable conditions of international competition—not to mention the 'Faustian' character of human nature, which drives men to increase their power over nature to the point of substituting built-in obsolescence for the satisfaction of fundamental needs—makes it impossible for knowledge to attain its ideal ends without continuing to serve the ends of power.

In this sense, science policies provide left-wing criticism with ideal material for the indictment it presents against industrial societies; in technonature, the ends of knowledge finish by coinciding so perfectly with the ends of power that scientific work seems, as it were, fated to assume the forms of unreason. What is surprising, moreover, is not that there should be a revolt against this use of science, but that this revolt should have taken so long to spread to the masses. The indictment of Marcuse is new only to those who have not learned the pessimistic lessons of Spengler, of Husserl, or even of Weber, each of them denouncing, in the terms of his own special preoccupations, the illusions cherished by the inheritors of the eighteenth century; the rationality of the scientific process, far from interrupting the irrationality of history, merely adds to it. Perhaps we had to arrive at the era of research organised in the light of political decisions for science to lose beyond recovery its childhood innocence and the enchantment of its promises.

Other choices, other priorities, would be possible for the policies whose subject matter is science. If this is a dream, we must confess that industrial societies leave little room for dreams; men may well ask what scientific and technical progress will make of them, while they have forgotten what they could make of scientific and technical progress. The effect of science policies, if they will persist in thus reducing knowledge to its instrumental function, is inevitably to inspire nihilism. If science is the subject of a revolt—not, as Spengler prophesied, 'the mutiny of the Hands against their destiny',[18] but the mutiny of the spirit against the ends it serves—it is not because science gives too little, but because it gives too much. Scepticism was all very well for science before Bacon, which had not proved itself; the modern version of this scepticism questions not what science *is* but what it *can do*. The association of knowledge and power, as Bacon saw it, was to legitimise modern science; technonature has above all legitimised the modern state as the leading producer and consumer of knowledge—which must be interpreted as *oriented* knowledge.

It may be that any other orientation should be ruled out because human nature is incapable of learning how to reverse the relationship of knowledge and power and to make the absurd ends of power yield to the ideal ends of knowledge. Science is a procedure of reason only within its own limits and it is hard to see what force in the world could enable it to subordinate the language of history. The illusion, on the contrary, is to expect the application of science to human affairs to provide ready-made solutions for problems whose key is fundamentally political. The 'first generation' of science policies developed after the second world war gave preference to the natural sciences; if the second generation proves to be that of the social sciences, there is a fear that by being recognised in their turn for their instrumentality alone, they will be summoned to the rescue as agents for the maintenance of the existing order and not primarily to question the meaning of that order.

Precisely on the pretext of the social traumas provoked by the speeding up of technical change, the social sciences would then prolong in the name of ideology the function which technonature has assigned to the natural sciences in the name of their scienticity: the manipulation of man following on the manipulation of things. Even if the conflicts of value which are inevitable in the use of any science are more explicit in the social sciences than in the natural sciences they will not therefore be resolved on the basis of technical criteria. In the absence of an established and shared conception of the 'health' of the body social or the 'quality' of the environment, this function of healing the evils of the technological epidemic which is proposed for the social sciences is hardly likely to be anything more than a police function.

It is not evident that modern societies can ever overcome the contradiction they have forged between the 'active agent' of culture embodied in the scientific spirit and method and the 'humanist' element repre-

18. O. Spengler, *Man and Technics* (trans. Charles Francis Atkinson), Knopf, New York (1932), 98.

sented by those who profess no practical skill. The 'two cultures' are a
drama, not as Lord Snow thinks [19] because they respond to two opposing orders of tastes, interests or even skills, but because they bring into
opposition functions which the industrial system holds out as irreconcilable; if the natural sciences train technicians, the social and human
sciences do not thereby train amateurs. What might well be called the
imperialism of scientific culture does not necessarily imply such a dichotomy between thought divorced from action and practice divorced from
reflection. The process of industrialisation gives preference in its scale
of values to skills and activities which most effectively meet its needs,
but no one can measure the weight carried in the social system by thinking, skills and labour which elude all calculation of economic return.

The conquests of science have been said to be limited to two spheres,
'organised simplicity' and 'disorganised complexity', the regularities of
the former being expressed in invariable laws and those of the latter in
statistical laws.[20] Now the world in which man is in future condemned
to live is that of *organised complexity*, to which the language, methods
and successes of the natural sciences are still only one key among others.
But this key works, and if it does not work to open other doors, that
does not mean that it turns idly in locks it was not made to fit. Over
and above the social difficulties it has not solved, or those which it has
itself created, one feels that the present revolt against science is aimed
not only against the lines it has chosen to follow in technonature, but
against the scientific attitude itself. In this sense, radical criticism
tends, confusedly or explicitly, to confound the misdeeds of a social
practice of science subordinated to the imperatives of the industrial
system and international competition with some evil inherent in the
nature of the scientific approach. The common characteristic of all these
forms of revolt is that they appeal to an experience which eludes the
influence of scientific thought as such, and of the 'technocratic' procedures in which it is embodied: a different experience of a different
reality, founded no longer on the final authority of proof and experiment, but on the non-intellectual or 'an-intellectual' values of instinct
and of life.[21]

The 'unreasonable' element in technonature, in so far as the ends of

19. C. P. Snow, *The Two Cultures*, Cambridge University Press (1959).
20. Warren Weaver, 'Science and Complexity', *American Scientist*, 36, no. 4 (1948).
21. A theme developed, for example, in the recent book by Theodore Roszak, *The Making of a Counter Culture: Reflections on the Technocratic Society and its Youthful Opposition*, Doubleday, New York (1969); Faber, London (1971).

knowledge are subordinated to the ends of power, leads some people to repudiate the whole rationality of science in the name of a different reason presiding over the happiness of a 'pacified' society, liberated from the restraints of complexity and organisation. It is always possible to denounce, in the style of Rousseau, everything that man has lost by his intoxication with the sciences and technologies, but there is no escape, because there is no substitute for rational thought, unless of course we are prepared to give up all attempt to cope with the difficulties and tensions generated by organised complexity. Science and technology are our destiny, just as politics is the destiny of science and technology; we can learn to make better use of them, but we cannot escape from them.

III

There will always be 'savants', researchers who are not prepared to be diminished to mere technicians, but they will be fewer and fewer in the growing population of scientists. In an age of organised and industrialised research such a distinction is meaningless if the criterion is the practice of research, but it still has a meaning from the ideological point of view. In the past, the savant was contrasted primarily with the ignorant man, the believer in fables and superstition, the representatives of non-science or false science, in so far as he submitted to the imperatives of experiment and verification proper to modern science. But he not only embodied the legitimacy of science as dealing with truth; he was a savant, in the first place because he *knew*, but also because he *practised* science as a system of knowledge aimed at an end other than the mere practice of a method. Today the savant is still defined in our eyes by his ability to transcend technical knowledge in an intellectual and moral commitment aiming at something more than the mastery of a speciality; in this he is distinguished not only from the man who does not know, but also from the technician who knows without thereby 'practising' science with the same consciousness of its values and its impact on the world.

The scientist, on the other hand, *is made* by the practice of a speciality. He is distinguished from the technician in that he is not content to apply the results of research, but the ideology exemplified by his research practice is no different from that of a technician; in his eyes science is merely a method and a means without any other end. The scientist is defined professionally, the savant intellectually: the idea of

science is related to the savant as an individual identified by his vocation, the reality of *the sciences* is related to the scientist as a member of
a group identified by skills. Among researchers there are hundreds of
thousands of scientists, but there are, and can be, only a handful of
savants. The professionalisation of research was the first step towards
reducing the number of savants in the total population of researchers,
since training for a research career involved increasing specialisation
and at the same time a narrowing of the cultural horizon; the second
and much more decisive step has been the multiplication of needs for
scientific skills as a result of the speeding-up of the industrialisation
process over the last thirty years.

Scientists have found their identity as technicians of knowledge in
relation to the 'intellectual' professions based on the practice and ideology of other forms of knowledge, but it is by no means certain that
this recognition will be accompanied in future by the same promises of
status, prestige and openings for research that they have recently enjoyed thanks to the 'exponential' growth of the research system. To take
the example of the United States, the proportion of scientists to total
population has risen from less than 0.5 per cent in 1940 to more than
1 per cent; a simple projection of this growth curve suggests that the
total population of the United States will finally consist of scientists,
just as the projection of the growth rate of resources allocated to research and development in relation to gross national product indicates
that research and development may one day swallow up the whole
gross national product. Whatever the needs of 'post-industrial societies'
for scientific skills, it is clear that this growth rate cannot continue in-
definitely. If a balance is to be struck between the growth of research
expenditure and the growth of GNP, the expansion of the universities
and the growth of the student body should, on the contrary, tend to keep
up the 'output' of scientists at a faster rate than the growth of research
expenditure. Not all these scientists will, of course, become researchers,
but at least the great majority of those who obtain their doctorates will
take up research work.

We can therefore imagine a new state of the research system in which
the supply of scientific manpower largely exceeds the demand, even in
the most highly industrialised societies; apart from the PhDs, who can
always teach in the universities, the possibilities of a scientific career
will be slight for all the others. These tendencies, which are beginning to
show themselves in the United States, will be visible in all industrialised
countries: up to the present, scientific skills have been multiplied to

meet the needs created by the rapid growth of the research system
rather than to fill the natural wastage by death, retirement and change of
occupation. If the growth rate of research activities levels off, and even
more if it falls, the 'age pyramid' of the research population will be
identical with that of manpower in general. Thus there are two reasons
to speculate about the future, as shown by a specialist in questions of
scientific manpower: in so far as the 'creativity' of researchers is related
to their relative youth, a decline in the 'rate' of discovery, invention
and innovation is conceivable in proportion to the number of scientists
engaged in research; even more certainly, we may expect less obvious
openings and slower promotion possibilities for young researchers,
while the average age of department chiefs or laboratory directors rises
steadily in a network of research institutions condemned, if not to
diminish, at least to arrest its growth.[22]

It is not idle to put these questions to discover that the career and
function of scientists are not certain to be the same tomorrow as they are
today or, in any event, to meet the hopes of the new generation destined
for research; more and more scientists trained in and for research will
find themselves unable to follow the research career to which they
aspire. The professionalisation of scientific activities has transformed research into work which falls, like any other work, into the context of a
career, but research as such is no longer a lifelong career. The over-
production of scientists preparing for this career will merely 'swell the
host of intellectuals' spoken of by Schumpeter, who, as a result of being
unsatisfactorily employed or unemployable, drift into other vocations
'in a thoroughly discontented frame of mind'.[23]

The professionalisation of scientists would thus lead to their prole-
tarianisation; the fact that these new proletarians, in spite of their
technical qualifications, in their turn swell the ranks of 'drop-outs' of
the 'affluent society' will not be one of the least paradoxes of an industrial system which incessantly calls for higher and higher technical
qualifications. This means at the very least that the promised land of
the 'post-industrial society', in which each worker is called upon to be an
intellectual artificer of the industry of knowledge, will not exactly meet
the wishes of most scientists who look to it for a different kind of social
betterment and social recognition. But, precisely, the scientist is an

22. See Joseph P. Martino, 'Science and Society in Equilibrium', *Science*, **165**
(22 August 1969), 769–72.
23. Joseph A. Schumpeter, *Capitalism, Socialism and Democracy* (3rd edition), Allen and Unwin, London (1950), 153.

intellectual only by virtue of the function he fulfils, where the savant is an intellectual by virtue of the vocation which defines him; the scientist's function may take the place of a vocation for him, while only the savant's vocation takes account of the function he fulfils. The most prestigious or the most attractive of the 'intellectual' professions of the industrial system is not necessarily destined to remain so.

The scientific research system, which consecrates the industrialisation of research work after having professionalised it, can very well do without the savant as such; all it needs is technicians. The line of demarcation between savants and scientists is traced not by the researcher's motives, his aptitude for research, or even by the more or less application-oriented character of his research work; the scientist may be just as much a 'great researcher', just as committed to 'fundamental' work, moved by just the same disinterested motives as the savant. But while the science which they practise is the same, the science which they think, live and *act* is not; the scientist expresses a professional reality, the savant an ideological imperative. The idea of the savant stems from a conception of research which rejects everything that it has become as a productive activity in our industrialised systems in favour of what it tended to be as an ideology of knowledge in the pre-industrial systems.

A nostalgic idea, no doubt, which looks back to an age of culture in which knowledge embodied in a skill did not conflict with knowledge assumed in all its intellectual and moral imperatives. But this nostalgia for a scientific institution different from what it is does not rule out individual behaviour defined by what it was; if industrialisation turns research into a job like any other, not all researchers can resign themselves to being no more than the executants of a science without soul or conscience. The scientist is a *production agent*; the savant is *a culture figure*. As Auguste Comte emphasised in denouncing in advance a positivism which he wished to distinguish from his own, what discriminates the savant from the scientist is 'the predominance he accords to the spirit of the whole over the spirit of detail', or, in other words, his desire to transcend the purely functional aspect of technical knowledge. The savant looks beyond technical knowledge and sees science both as an imperative of unity and as a clearing-house of values; his approach to his research work is not that of utility but that of an end which transcends his technical mastery of a given and necessarily limited sector of scientific inquiry. Even if he is financed by a private institution, the scientist's work is the work of a functionary; the savant, in contrast, is closer to the philosopher, in the sense in which Husserl defined him, in his work itself, as a 'functionary of mankind',[24] a man for whom research goes hand in hand with the sense of a responsibility for ends that go beyond the purely technical field of his professional activity.

It is the savant who speaks in the scientist when he resists the alienation of technonature and questions the ends served by the knowledge of which he holds the monopoly; the ideology of science conceived as the sanctuary of an objective, neutral procedure, sheltered from mundane speech and the compromises of politics, does not restrain him from taking part in the affairs of this world. If his intervention then no longer appears to be that of one citizen among others, whose greater skill is never any guarantee of truth, it is nevertheless in the name of science conceived as the search for truth that he questions the meaning and political implications of the scienticity of his profession. In establishing its legitimacy in the culture system, modern science was at the same time bound to proclaim and defend the autonomy of its approach against the magisterial and overbearing spirit displayed by all authority, religious, political or economic. Once this autonomy is won, it is true that it can be challenged at any moment. But autonomy is not neutrality; for all that neutrality is erected as a barrier to protect autonomy, even at the cost of perpetuating the ancient myth of thought divorced from action and knowledge indifferent to power, science is all the less sheltered from the tempests of history in that it is no stranger to unleashing them.

So long as they are supported by the community, the sciences will continue to multiply and fashion technical aptitudes and skills of this type proper to the scientist—the technician of the natural sciences, the possessor of knowledge to which the uninitiated has no access and which, in any event, he has no intention of sharing lest he should find his social function 'demystified' and clearly recognised as merely one function among others, an anonymous and therefore less prestigious part of mass research. But there will always be scientists, however rare they are becoming, who will behave like savants or, in other words, like researchers anxious to overcome the tendency for their technical mastery to subordinate their conception of research to the wishes of power, to enlarge their practice of a speciality to a deeper understanding of their social role and consequently throw this understanding open to the non-specialist, to inform him and influence him, albeit through the

24. E. Husserl, *The Crisis of European Sciences and Transcendental Phenomenology* (trans. David Carr), Northwestern University Press, Evanston, Illinois (1970), 17.

ambivalence of political commitment: the savant is not only the man who knows and practises science for science's sake, he is also the man who thinks of it as a problem for mankind. The sciences are the affair of an army of technicians indifferent to the values which science challenges in its relations with the polity. Science, if it is to be preserved as a realm of values rather than an intellectual technique, can be preserved only by individuals.

Postscript

Since this book was first published in France, in September 1970, many other books on the same or kindred subjects have been published in many parts of the world. The subject lends itself to fresh analysis, not just because there is such a wealth of material but mainly because there is a steadily growing awareness of the weight of scientific activities in world affairs. At the same time, people are recognising the danger of allowing a free run to an economic system which aims at growth for growth's sake heedless of the resulting social costs and inequalities, and to science policies which rely on 'gadgets'—in the case of states, mega-gadgets generated by competition between the great powers, the struggle for prestige or the conquest of military markets, and in the case of industry, mini-gadgets generated by the exclusive hunt for innovation and profit—heedless of the real needs of mankind.

The purpose of my book, among others, was to denounce this misuse of science and technology—of technology *and* of science—which can be given a rational guise precisely by resorting to scientific methods, techniques and language, and under this rational guise to denounce the complicity of most scientists, their ambiguous function and their *de facto* alienation from political power. The indictment has unquestionably gathered force. But while I feel bound to mention the books and reports published since 1970 which throw light on the debate, whether or not they constitute an extension of certain aspects of my own analysis, I find nothing to change today either in the context of this analysis or in the analysis itself, or even in the examples and data adduced in support. I will cite one example only. Chapter 5, on technological forecasting and the myths it has encouraged in the matter of science policy, draws partly on E. Jantsch's book published in 1967, and this is no accident: it was the first major work on the subject. Since that date a mass of books have been published on technological forecasting and its social and political extensions, including one by Jantsch himself: *Perspectives of Planning*, OECD, Paris (1971). Everything I wrote then to show the limits and illusions of this method I would say again today in the same language, with the benefit of other works and sometimes, as in this specific instance, of the experience acquired by specialists . . . who have now become experts: see *Analytical Methods in Government Science Policy—An Evaluation*, OECD, Paris (1972). Rather than burden this book with additional footnotes on the most recent books in response to the very academic temptation of emphasising the fact that they say the same thing as mine or say it in similar terms—better or less well chosen—it seems to me both simpler and more honest to cite here those books which, published after mine, seem to me to contribute either new information or an original approach to the problems I have discussed, and to cite them in alphabetical order, to avoid hurting the susceptibilities of any of the authors concerned—including myself!

BEN-DAVID, Joseph, *The Scientist's Role in Society: a comparative study*, Prentice Hall, New Jersey (1970).

BROWN, Martin (ed.), *The Social Responsibility of the Scientist*, The Free Press, New York (1971).

CAHN, Anne Hessing, *Eggheads and Warheads: Scientists and the ABM*, PhD Dissertation (stencilled), Department of Political Science, MIT, Cambridge, Mass. (1971).

CALDWELL, Linton K., *In Defense of Earth*, Indiana University Press, Bloomington (1972).

CATY, G., DRILHON, G., FERNE, G.,WALD, S. (under the direction of SALOMON, J.-J.), *The Research System—Comparative Survey of the Organisation and Financing of Fundamental Research*, volume 1: *France, Germany, United Kingdom*, OECD, Paris (1971); volume 2: *Belgium, Netherlands, Norway, Sweden, Switzerland* (1973).

CLARKE, Robin, *We All Fall Down*, Penguin, Harmondsworth (1969).

CRANE, Diana, *Invisible Colleges*, University of Chicago Press (1972).

DOBROV, G. M., *Wissenschafts-Organisation und Effectivität*, Akademie-Verlag, Berlin (1971) (translated from Russian).
    Potential der Wissenschaft, Akademie-Verlag, Berlin (1971) (translated from Russian).

FELD, B. T., GREENWOOD, R., RATHJENS, G. W., and WEINBERG, S. (ed.), *Impact of New Technologies on the Arms Race*, MIT Press (1971).

FORRESTER, Jay W., *World Dynamics*, Wright-Allen Press, Cambridge, Mass. (1971).

JAUBERT, A., and LEVY-LEBLOND, J. M., *(Auto)critique de la Science*, Seuil, Paris (1973).

MAUNOURY, Jean-Louis, *Economie du Savoir*, A. Colin, Paris (1972).

MEADOWS, Dennis L., *The Limits to Growth*, Wright-Allen Press, Cambridge, Mass. (1972).

MEDVEDEV, Zhores, *The Rise and Fall of T. D. Lysenko*, Columbia University Press (1969).

MIT (sponsor), *Man's Impact on the Global Environment*, MIT Press, Cambridge, Mass. (1970).

MUMFORD, Lewis, *The Myth of the Machine—The Pentagon of Power*, Secker and Warburg, London (1971).

NELKIN, Dorothy, *The University and Military Research—Moral Politics at MIT*, Cornell University Press, Ithaca (1972).

OECD, *Science, Growth and Society—A New Perspective*, report of the Secretary General's *ad hoc* group on the new concepts of science policy (the 'Brooks Report'), OECD, Paris (1971).

RAVETZ, Jerome R., *Scientific Knowledge and its Social Problems*, Clarendon Press, Oxford (1971).

ROSE, Hilary, and ROSE, Steven, *Science and Society*, Penguin Books, Harmondsworth (1970).

ROSZAK, Theodore, *Where the Wasteland Ends*, Doubleday, New York (1972).

SAPOLSKY, Harvey M., *Creating the Invulnerable Deterrent—Progammatic and Bureaucratic Success in the Polaris System Development*, Harvard University Press, Cambridge, Mass. (1972).

SCHATZMAN, Evry, *Science et Société*, Laffont, Paris (1971).

SCHOOLER, Dean, Jr., *Science, Scientists and Public Policy*, The Free Press, New York (1971).

SKOLINIKOFF, Eugene B., *The International Imperatives of Technology—Technological Development and the International Political System*, Research Series no. 16, Institute of International Studies, University of California, Berkeley (1972).

WEINGART, Peter, *Die Amerikanische Wissenschaftslobby*, Bertelsmann Universitätstering, Düsseldorf (1970).

WEISSKOPF, Victor F., *Physics in the Twentieth Century*, MIT Press, Cambridge, Mass. (1972).

Review articles are obviously countless, and I refrain from mentioning any except the following which are of outstanding relevance or importance:

'Blueprint for Survival', *The Ecologist*, 2, no. 1 (January 1972).

FREEMAN, Christopher, *et al.*, 'The Goals of R & D in the 1970s', *Science Studies*, 1, nos. 3/4 (1971), 357–406.

JOHNSON, Harry G., 'Some Economic Aspects of Science', *Minerva* (January 1972), 10–18.

MANSFIELD, Edwin, 'R & D's Contribution to the Economic Growth of the Nation', *Research Management* (May 1972), 31–46.

'Science et Société', a colloquium organised jointly by the French DGRST and OECD at Saint-Paul de Vence, *Le Progrès scientifique*, no. 160, DGRST, Paris (January 1973).

SERRES, Michel, 'La Thanatocratie', *Critique* (March 1972), 199–227.

'World's Dynamics Challenged' (to be published as *Thinking about the Future* by Sussex University Press), a critical analysis of *The Limits to Growth*, *Futures*, 5, no. 1 (February 1973) and no. 2 (April 1973), IPC House, Guildford, Surrey, England.

Bibliography

A complete bibliography on the relations between science and society is by definition an impossible undertaking, worthy of Bouvard and Pécuchet: any effort in this field is doomed to be partial and, in a word, arbitrary. No doubt the social, economic and political aspects of science have only been the subject of specific research, except of course by historians proper, for the last fifty years or so, but the development of research activities and their social repercussions within the last twenty-five years or less have swollen world literature in this field to such a point that, even on particular aspects of the issues at stake, it becomes constantly more difficult to take stock, particularly if it is desired to include articles in periodicals as well as books and documents.

As long ago as 1936 the sociologist Ossowski and his wife advocated a 'science of science' to study everything that makes science a social institution linked to society as a whole;[1] the object of such a discipline seems indefinite, and it still has to be constructed from scratch, but there does in fact exist today a sociology and an economics of science as well as a wealth of political science studies connected with the development of science and technology. In this literature on the relations between science and society, the works more specifically devoted to science policies form a corpus of their own in that they are easier to identify (as well as being in greater vogue), but, like the visible part of an iceberg, they still display only the more obvious portion of a field of research and reflection whose unexplored area is certainly far more vast.

In any event there is now in existence a remarkable bibliography on the subjects discussed, compiled under the direction of Lynton K. Caldwell, and in which the present author collaborated: *Science, Technology and Public Policy: A Selected and Annotated Bibliography*, vol. 1 (books, monographs, government documents and special numbers of periodicals, 492 pp., 1968) and vol. 2 (articles in periodicals, 544 pp., 1969), Indiana University, Bloomington, United States.[2] To this must

1. Maria Ossowska and Stanislas Ossowski, 'The Science of Science', *Organon*, Warsaw, 1 (1936), 1–12.
2. A new volume for 1945–67 includes both books and articles (868 pp., 1973).

be added the bibliography compiled by Norman Kaplan in his *Science and Society*, Rand McNally and Co., Chicago (1965), 581–95, and that given at the end of *Problems of Science Policy*, OECD, Paris (1968), 193–215.

Referring the reader to these bibliographies, and especially to the first, which has the advantage not only of being more complete but also of being annotated, we content ourselves with noting here some of the books cited as references in the course of our essay, omitting articles but including some other texts which have given us food for thought.

Mention should be made of the recent efforts to reprint essays and articles on the social aspects of science originally published in the nineteenth century, in the collection *Science and Public Affairs* (ed. Roy M. MacLeod), Cass, London. Finally, it may be noted that the main periodicals dealing with relations between science and politics are *Nature, New Scientist, Minerva* and, since 1971, *Science Studies*, in Great Britain; *Science, Scientific American* and *The Bulletin of the Atomic Scientists* in the United States; and *La Recherche* (formerly *Atomes*) in France. Three special numbers of periodicals are worth consulting: 'Science and Culture', *Daedalus*, Cambridge, Mass. (winter 1965); 'La Recherche scientifique, l'Etat et la Société', *Prospective*, no. 12 (1965); and 'Science et politique', *Les Etudes philosophiques*, no. 2 (April–June 1966), P.U.F., Paris.

AILLERET, Charles, *L'Aventure atomique française*, Grasset, Paris (1968).
ARENDT, Hannah, *The Human Condition*, University of Chicago Press, Chicago (1958).
    Between Past and Future, Viking Press, New York (1968).
ARON, Raymond, *La Société industrielle et la Guerre*, Plon, Paris (1952).
    Peace and War – A Theory of International Relations, New York (1966).
    Le grand débat, Calmann-Lévy, Paris (1963).
    Democracy and Totalitarianism (trans. Valence Ionescu), Weidenfeld and Nicolson, London (1968).
    Trois essais sur l'age industriel, Plon, Paris (1966).
    Main Currents in Sociological Thought (trans. Richard Howard and Helen Weaver), Weidenfeld and Nicolson, London (1968).
AUGER, Pierre, *Current Trends in Scientific Knowledge*, Unesco, Paris (1961).
BACHELARD, Gaston, *La formation de l'esprit scientifique*, Vrin, Paris (1938).
    La philosophie du non, P.U.F., Paris (1940).
    Le rationalisme appliqué, P.U.F., Paris (1949).
    L'activité rationaliste de la physique contemporaine, P.U.F., Paris (1951; 2nd edition, 1965).
    Le matérialisme rationnel, P.U.F., Paris (1953).
BACON, Francis, *The New Atlantis* (ed. A. B. Gough), Oxford (1924).
    Advancement of Learning (ed. G. W. Kitchin), London (1950).
BARBER, Bernard, *Science and the Social Order*, The Free Press, New York (1952).

The Sociology of Science, The Free Press, New York (1962).

BARGHOORN, F. C., *The Soviet Cultural Offensive*, Princeton University Press, New Jersey (1960).

BAUER, Raymond A., *Social Indicators*, MIT Press, Cambridge, Mass. (1967).

BEN-DAVID, Joseph, *Fundamental Research and the Universities*, OECD, Paris (1967).

BERNAL, J. D., *The Social Function of Science*, Routledge and Kegan Paul, London (1939; reissued by MIT Press, 1967).

Marx and Science, International Publishers, New York (1952).

Science and Industry in the Nineteenth Century, Routledge and Kegan Paul, London (1953).

Science in History (1954; reissued by Penguin, Harmondsworth, 4 volumes, 1969).

BORN, Max, *La responsabilité du savant*, Payot, Paris (1967).

BRONOWSKI, Jacob, *Science and Human Values*, Messner, New York (1958); revised edition, Penguin, Harmondsworth (1964).

BROOKS, Harvey, *The Government of Science*, MIT Press, Cambridge, Mass. (1968).

BUSH, Vannevar, *Science, The Endless Frontier: A Report to the President on a Program for Postwar Scientific Research* (1945; reissued by the National Science Foundation, Washington, 1960).

CALDER, Nigel, *Technopolis: social control of the uses of science*, MacGibbon and Kee, London (1969; reissued by Panther Books, 1970).

CANDOLLE, Alphonse de, *Histoire des sciences et des savants depuis deux siècles*, Geneva–Basle (1873).

CANGUILHEM, Georges, *Etudes d'histoire et de philosophie des Sciences*, Vrin, Paris (1968).

CASSIRER, Ernst, *The Philosophy of the Enlightenment* (trans. F. C. A. Koelln and J. P. Pettegrove), Princeton University Press (1969).

CHAYES, Abraham, and WIESNER, Jerome B., *ABM: An Evaluation of the Decision to Deploy an Antiballistic System*, New American Library, New York (1969).

CHOMSKY, Noam, *American Power and the New Mandarins*, Pantheon Books (1969) and Penguin (1969).

CLAUSEWITZ, K. von, *On War*, Random House, New York (1943).

CLOSETS, François de, *En danger de progrès*, Donoël, Paris (1970).

COLETTE, J. M., *Recherche–Développment et progrès économique en URSS*, Cahiers de l'ISEA, Paris (1962).

COMMONER, Barry, *Science and Survival*, Viking Press, New York (1967).

COMPTON, Arthur H., *Atomic Quest: A Personal Narrative*, Oxford University Press, New York (1956).

COMTE, Auguste, *Cours de Philosophie positive* (ed. Schleicher Frères), Paris 1907–8).

Système de politique positive, commemorative edition by the Société positiviste, Paris (1942).

CONANT, James B., *Modern Science and Modern Man*, Doubleday, New York (1952).

CONDORCET, *Eloge des académiciens à l'Académie royale des Sciences* (1773 edition).

Esquisse d'un tableau historique des progrès de l'esprit humain, in *Oeuvres complètes* (ed. Garat–Cabanis), Paris (1804).

CROMBIE, A. C. (ed.), *Scientific Change: Historical Studies in the Intellectual*,

Social and Technical Conditions for Scientific Discovery and Technical Invention, Oxford (1961), Basic Books, New York.

CROWTHER, J. G., *Statesmen of Science*, Cresset Press, London (1952).
    Francis Bacon: The First Statesman of Science, Cresset Press, London (1960).

CROZIER, Michel, *Le Phénomène bureaucratique*, Editions du Seuil, Paris (1963).

DAGOGNET, François, *Méthodes et doctrine dans l'oeuvre de Pasteur*, P.U.F., Paris (1967).

DALMAS, André, *Evariste Galois*, Fasquelle, Paris (1956).

DAUMAS, Maurice, *Les instruments scientifiques aux XVIIe et XVIIIe siècles*, Paris (1953).
    (general editor) *Histoire générale des techniques*, 3 volumes published, P.U.F., Paris (1962–9).

DAVIS, Nuel Pharr, *Lawrence and Oppenheimer*, Simon and Schuster, New York (1968).

DE BEER, Gavin, *The Sciences were Never at War*, Nelson, London (1960).

DENISON, Edward F., *The Sources of Economic Growth in the United States and the Alternatives before us*, Committee for Economic Development, New York (1962).
    Why Growth Rates Differ: postwar experience in nine western countries, The Brookings Institution, Washington (1967).

DOBROV, G. M., *Nauka o Nauka (The Science of Science: Introduction to the General Study of Science)*, Kiev, Naukova Dumka (1966).

DUBARLE, Dominique, O.P., *La civilisation et l'atome*, Editions du Cerf–Plon, Paris (1962).

DUBOS, René J., *The Dreams of Reason: Science and Utopias*, Columbia University Press (1961).

DUPRE, J. Stefan, and LAKOFF, Sanford A., *Science and the Nation: Policy and Politics*, Prentice-Hall, N.J. (1962).

DUPREE, A. Hunter, *Science in the Federal Government*, Harper Torchbook, New York (1964).

EASTON, David, *The Political System—An Inquiry into the State of Political Science*, Knopf, New York (1960).

EINSTEIN, Albert, *Comment je vois le monde*, Flammarion, Paris (1934).
    Conceptions scientifiques, morales et sociales, Flammarion, Paris (1952).

ELLUL, Jacques, *The Technological Society*, Cape, London (1965).

FARRINGTON, Benjamin, *Science and Politics in the Ancient World*, George Allen and Unwin, London (1965).

FAYET, Joseph, *La Révolution française et la science*, Marcel Rivière, Paris (1960).

FICHTE, J. G., *La destination du savant* (1794), Vrin, Paris (1969).

FLEMING, D., and BAILYN, B. (ed.), *The Intellectual Migration*, Harvard University Press (1968).

FREEMAN, Christopher, and YOUNG, Alison, *The Research and Development Effort in Western Europe, North America and the Soviet Union*, OECD, Paris (1965).

FRUTKIN, Arnold W., *International Cooperation in Space*, Prentice-Hall, New Jersey (1965).

GALBRAITH, J. K., *The New Industrial State*, Houghton Mifflin, Boston (1967).

GILFILLAN, S. C., *The Sociology of Invention*, Follet Publishing Co., Chicago (1935).

GILLISPIE, Charles Coulston, *The Edge of Objectivity—An Essay in the History of Scientific Ideas*, Princeton University Press (1960).

GILPIN, Robert, *American Scientists and Nuclear Weapons Policy*, Princeton University Press, New Jersey (1962).
    France in the Age of the Scientific State, Princeton University Press (1968).

GILPIN, Robert, and WRIGHT, Christopher (ed.), *Scientists and National Policy-Making*, Columbia University Press, New York and London (1964).

GLUCKSMANN, André, *Le discours de la guerre*, Editions de l'Herne, Paris (1967).

GOLDSCHMIDT, Bertrand, *Les rivalités atomiques 1939–1966*, Fayard, Paris (1967).

GOLDSMITH, Maurice, and MACKAY, Alan (ed.), *The Science of Science—Society in the Technological Age*, Souvenir Press, London (1964).

GOLDSMITH, Maurice, and DE REUCK, Anthony (ed.), *Decision-making in National Science Policy*, a CIBA Foundation and Science of Science Foundation Symposium, Churchill, London (1968).

GOUDSMIT, Samuel A., *Alsos: The Failure of German Science*, Henry Schuman, New York (1947).

GOWING, Margaret, *Dossier secret des relations atomiques entre Alliés (1939–1945)*, Plon, Paris (1965).

GRAHAM, Loren A., *The Soviet Academy of Sciences and the Communist Party*, Princeton University Press (1967).

GREENBERG, Daniel S., *The Politics of Pure Science*, New American Library, New York (1967); published in England as *The Politics of American Science*, Penguin, Harmondsworth (1969).

GRODZINS, M., and RABINOWITCH, E. (ed.), *The Atomic Age*, Simon and Schuster, New York (1965).

HABERMAS, Jürgen, *Toward a Rational Society*, Heinemann, London (1971).

HAGSTROM, Warren O., *The Scienfic Community*, Basic Books, New York (1965).

HAAS, Ernst B., *Beyond the Nation State*, Stanford University Press, California (1965).

HELMER, Olaf, *Social Technology*, Basic Books, New York (1966).

HEWLETT, R. G., and ANDERSON, O. E., Jr., *The New World—A History of the United States Atomic Energy Commission*, vol. 1, Pennsylvania State University Press (1962).

HEWLETT, R. G., and DUNCAN, F., *Atomic Shield 1947–1952: A History of the United States Atomic Energy Commission*, vol. 2, Pennsylvania State University Press (1969).

HILL, Karl (ed.), *The Management of Scientists*, Beacon Press, Boston (1964).

HITH, Charles J., and McKEAN, Roland M., *The Economics of Defense in the Nuclear Age*, Harvard University Press (1960; reissued by Athenaeum, New York, 1965).

HOFFMANN, Stanley, *Gulliver's Troubles, or the Setting of American Foreign Policy*, McGraw–Hill, New York (1968).

HUIZINGA, Johan, *Homo Ludens: A Study of the Play-Element in Culture*, Routledge and Kegan Paul, London (1949).

HUSSERL, Edmond, *The Crisis of European Science and Transcendental Phenomenology*, Northwestern University Press, Evanston, Illinois (1970).

JANTSCH, Erich, *Technological Forecasting in Perspective*, OECD, Paris (1967).

JASPERS, Karl, *La bombe atomique et l'avenir de l'homme*, Buchet-Chastel, Paris (1963).

JEWKES, J., SAWERS, D., and STILLERMAN, R., *The Sources of Invention*, Macmillan, London (1969).

JORAVSKY, David, *Soviet Marxism and Natural Sciences*, Columbia University Press (1961).

JOUVENEL, Bertrand de, *L'art de la conjecture*, Editions du Rocher, Monaco (1964).

JUNGK, Robert, *Brighter Than a Thousand Suns* (trans. James Clough), Gollancz and Hart-Davis (1958; reissued by Penguin, Harmondsworth, 1960).

KAHN, Herman, *On Thermonuclear War*, Princeton University Press, New Jersey (1961).

    De l'escalade—métaphores et scenarios, Calmann-Lévy, Paris (1966).

KAHN, Herman, and WIENER, Anthony J., *L'An 2000*, Laffont, Paris (1968).

KANT, Immanuel, *Critique of Pure Reason* (ed. Norman Kemp Smith), Macmillan, London (1958).

KAPITZA, P. L., *Peter Kapitza on Life and Science*, Macmillan, New York (1968).

KAPLAN, Norman (ed.), *Science and Society*, Rand, McNally and Co, Chicago (1965).

KERR, Clark, *The Uses of the University*, Harper Torchbooks, New York (1966).

KLAW, Spencer, *The New Brahmins: Scientific Life in America*, Morrow, New York (1968).

KOROL, Alexander G., *Soviet Research and Development: its Organisation, Personnel and Funds*, MIT Press, Cambridge, Mass. (1965).

KOURGANOFF, Vladimir, *La recherche scientifique*, P.U.F., Paris (1962).

KOYRÉ, Alexandre, *Etudes galiléennes*, Hermann, Paris (1939; 2nd edition, 1966).

    From the Closed World to the Infinite Universe, John Hopkins Press (1969).

    Etudes d'histoire de la pensée scientifique, P.U.F., Paris (1966).

KRANZBERG, Melvin, and PURSELL, Carroll W. (ed.), *Technology in Western Civilisation* (2 volumes), Oxford University Press, New York (1967).

KUHN, Thomas S., *The Structure of Scientific Revolutions*, Chicago University Press (1962; second edition, 1970).

KUZNETS, Simon, *Modern Economic Growth*, Yale University Press, New Haven (1966).

LAKOFF, Sanford A. (ed.), *Knowledge and Power*, The Free Press, New York (1966).

LAPP, Ralph E., *Kill and Overkill—The Strategy of Annihilation*, Basic Books, New York (1962).

    The New Priesthood: The Scientific Elite and the Use of Power, Harper and Row, New York (1965).

LASSWELL, Harold D., *The Analysis of Political Behaviour: An Empirical Approach*, Oxford University Press, New York (1948).

LEONTIEFF, W. W., *Studies in the Structure of the American Economy*, Oxford University Press, New York (1953).

LEWIN, Leonard C., *Report from Iron Mountain*, Macdonald, London (1968; reissued by Penguin, Harmondsworth, 1968).

MACHLUP, Fritz, *The Economics of Technological Change*, Norton and Co., New York (1968).

MARCUSE, Herbert, *One-Dimensional Man*, Beacon Press, Boston (1958), and Routledge, London (1964).

MARX, Karl, *Capital: A Critique of Political Economy*, Charles H. Kerr and Co., Chicago (1906, 1907, 1909).
    A Contribution to the Critique of Political Economy, Kegan Paul, London (1904).

MARX, Karl, and ENGELS, Friedrich, *The German Ideology*, International Publishers, New York (1939).
    Manifesto of the Communist Party (authorised English translation: edited and annotated by Friedrich Engels, translated by Samuel Moore), International Publishers, New York (1932); Martin Lawrence, London (1933).

MASSÉ, Pierre, *Le Plan ou l'anti-hasard*, Gallimard, Paris (1965).

MAUNOURY, Jean-Louis, *La genèse des innovations—La création technique dans l'activité de la firme*, P.U.F., Paris (1968).

MAUPERTUIS, G., *Lettres sur le progrès des sciences, Oeuvres*, Dresden (1752).

MERLEAU-PONTY, Maurice, *Eloge de la philosophie et autres essais*, Gallimard, Paris (1960).

MERTON, Robert K., *Science, Technology and Society in Seventeenth-Century England*, Bruges, Imprimerie Saint-Catherine (1938; reissued by Harper Torchbooks, New York, 1970).
    Social Theory and Social Structure, Free Press, New York (1957).
    Eléments de théorie et de méthode sociologique, Plon, Paris (1965).

MESTHENE, Emmanuel G. (ed.), *Ministers and Science*, OECD, Paris (1965).
    Technological Change: its impact on man and society (revised version of the Fourth Report of the *Program on Technology and Society*, Harvard University, Cambridge, Mass., 1967–8), Mentor Books, New York (1970).

MEYNAUD, Jean, *Technocracy* (trans. P. Barnes), Faber, London (1968).

MEYNAUD, Jean, and SCHROEDER, Brigitte, *Les savants dans la vie internationale*, Etudes de Science politique, Lausanne (1962).

MILL, John Stuart, *Utilitarianism*, Everyman's Library, Dent, London (1968).
    Principles of Political Economy, D. Appleton and Co., New York (1890); Penguin, Harmondsworth (1970).

MOONMAN, Eric (ed.), *Science and Technology in Europe*, Penguin, Harmondsworth (1968).

NATIONAL ACADEMY OF SCIENCES, Committee on Science and Public Affairs (COSPUP), Washington, *Basic Research and National Goals* (1965).
    Applied Science and Technological Progress (1967).
    Technology: Processes of Assessment and Choice (1969).

NELSON, Richard A., *The Rate and Direction of Inventive Activity*, Princeton University Press (1962).

NEUMANN, J. von, and MORGENSTEN, O., *Theory of Games and Economic Behaviour*, Princeton University Press, revised edition (1955).

NICHOLS, David, *The Political Attitudes of a Scientific Elite*, stencilled doctorate thesis, Department of Political Science, MIT, Cambridge, Mass. (1968).

NIEBURG, H. L., *In the Name of Science*, Quadrangle Books, Chicago (1966).

NOVICK, David, *Program Budgeting: Program Analysis and the Federal Budget*, Harvard University Press, Boston (1965).

OBLER, Paul C., and ESTRIN, A. Herman, *The New Scientist—Essays on the Methods and Values of Modern Science*, Doubleday, New York (1962).

OPPENHEIMER, J. Robert, *The Open Mind*, Simon and Schuster, New York (1955).
  In the matter of J. Robert Oppenheimer: A Transcript of Hearing before Personnel Security Board, and *Texts of Principal Documents and Letters* (2 volumes), USGPO, Washington (1954).
ORGANISATION FOR ECONOMIC COOPERATION AND DEVELOPMENT (OECD), *National Science Policies: Belgium* (1966), *France* (1966), *United Kingdom and Germany* (1967), *Japan* (1967), *United States* (1968), *Italy* (1969) and *Canada* (1969).
  Fundamental research and the universities; The social sciences and government policy; Governments and technical innovation; The allocation of resources to science (1966).
  International Statistical Year for Research and Development, vol. 1, *The Overall Level and Structure of R & D Efforts in OECD Member Countries* (1967); vol. 2, *Statistical Tables and Explanatory Notes* (1968).
  Reports on technological gaps: general report, plastics, pharmaceuticals, scientific instruments, electronic components, computers (1968–9).
  Science and Development: The experience of pilot teams (1968).
ORLANS, Harold (ed.), *Science Policy and the University*, The Brookings Institution, Washington (1968).
ORNSTEIN, Martha, *The Role of the Scientific Societies in the Seventeenth Century*, University of Chicago Press (1928).
OZBEKHAN, Hasan, *The Idea of a 'Look-Out' Institution*, Systems Development Corporation, Santa Monica, California (March 1965).
  Technology and Man's Future, Report SP-2494, Systems Development Corporation, Santa Monica, California (1966).
PASTEUR, Louis, *Oeuvres* (ed. P. Valléry-Radot) (vol. 8), Masson, Paris (1939).
PERROUX, François, *La coexistence pacifique* (3 volumes), P.U.F., Paris (1958).
  (general editor), *Recherche et activité économique*, Colin, Paris (1969).
PIGANIOL, Pierre, and VILLECOURT, Louis, *Pour une politique scientifique*, Flammarion, Paris (1963).
PIGANIOL, Pierre, *Maîtriser le progrès*, Laffont-Gonthier, Paris (1968).
POLANYI, Michael, *The Logic of Liberty*, Routledge and Kegan Paul, London (1951).
PRICE, Derek J. de Solla, *Science since Babylon*, Yale University Press, New York (1962).
  Little Science, Big Science, Columbia University Press, New York (1963).
PRICE, Don K., *Government and Science* (1953; reissued by Oxford University Press, 1962).
  The Scientific Estate, The Belknap Press of Harvard University, Cambridge, Mass. (1965).
RABINOWITCH, Eugene, *The Dawn of a New Age: Reflections on Science and Human Affairs*, University of Chicago Press (1963).
RENAN, Ernest, *L'Avenir de la science*, Calmann-Lévy (1890).
RICHTA, Rasovan, *La civilisation au carrefour—Implications sociales et humaines de la révolution scientifique et technique*, Editions Anthropos, Paris (1969).
ROE, Anne, *The Making of a Scientist*, Dodd, Mead & Co, New York (1952).
ROSTAND, Jean, *Science fausse et fausses sciences*, Gallimard, Paris (1958).
ROSZAK, Theodore, *The Making of a Counter Culture, Reflections on the*

Technocratic Society and its Youthful Opposition, Doubleday & Co., New York (1969); Faber, London (1971).

ROTBLAT, J., *Pugwash: A History of the Conferences on Science and World Affairs*, Czechoslovak Academy of Sciences, Prague (1967).

SAINT-SIMON, *Oeuvres* (vols. 4 and 5), Editions Anthropos, Paris (1967).

SAKHAROV, Andrei, *La liberté intellectuelle et la coexistence*, Gallimard, Paris (1965).

SALOMON, Jean-Jacques, *International Scientific Organisations*, OECD, Paris (1965).

(ed.), *Problems of Science Policy*, OECD, Paris (1968).

SANTILLANA, Giorgio de, *The Crime of Galileo*, Chicago (1955).

SCHELLING, Thomas C., *The Strategy of Conflict*, Galaxy Books, New York (1963).

SCHILLING, Warner, HAMMOND, Paul, and SNYDER, Glenn, *Strategy, Politics and Defense Budgets*, Columbia University Press, New York (1962).

SCHMOOKLER, Jacob, *Invention and Economic Growth*, Harvard University Press, Cambridge, Mass. (1966).

SCHON, Donald A., *Technology and Change*, Delta Books, New York (1967).

SCHUHL, P. M., *Machinisme et Philosophie*, Félix Alcan, Paris (1938).

SCHUMPETER, Joseph A., *Capitalism, Socialism and Democracy*, Allen and Unwin, London (1950).

SHILS, Edward, *The Torment of Secrecy: The Background and Consequences of American Security Policies*, The Free Press of Glencoe (1956).

(ed.), *Criteria for Scientific Development: Public Policy and National Goals*, MIT Press, Cambridge, Mass. (1966).

SHIMSHONI, Daniel, *Scientific Entrepreneurship*, stencilled doctorate thesis, Graduate School of Public Administration, Harvard University (1966).

SHONFIELD, Andrew, *Modern Capitalism*, Oxford University Press, New York (1965).

SIMON, Leslie E., *German Research in World War II*, John Wiley and Sons, New York (1947).

SIMONDON, Gilbert, *Du mode d'existence des objets techniques*, Aubier-Montaigne (1969).

SINGER, Charles, *et al.*, *A History of Technology* (5 vols.), Oxford (1958).

SIU, R. G. H., *The Tao of Science*, MIT Press, Cambridge, Mass (1964).

SKOLNIKOFF, Eugene B., *Science, Technology and American Foreign Policy*, MIT Press, Cambridge, Mass. (1967).

SMITH, Alice Kimbal, *A Peril and a Hope: The Scientists' Movement in America, 1945–1947*, The University of Chicago Press (1965).

SMITH, Bruce L. R., *The RAND Corporation*, Harvard University Press, Cambridge, Mass. (1966).

SNOW, C. P., *Science and Government* (1960; reissued by Mentor Books, New York (1962).

The Two Cultures, Cambridge University Press (1959).

SPAEY, Jacques (general editor), *Science for Development: an essay on the origin and organisation of national science policies*, Unesco, Paris (1971).

SPENGLER, Oswald, *The Decline of the West* (trans. C. F. Atkinson), Allen and Unwin (1932; abridged edition, 1961).

Man and Technics, Alfred A. Knopf, New York (1932).

STERN, Philip M., and GREEN, Harold P., *The Oppenheimer Case: Security on Trial*, Harper and Row, New York (1970).

STORER, Norman W., *The Social System of Science*, Holt, Rinehart and Winston, New York (1966).

SZILARD, Leo, *The Voice of the Dolphins*, Simon and Schuster, New York (1961).

TATON, René, *Reason and Chance in Scientific Discovery*, Philosophical Library, New York (1957).

(general editor), *A General History of the Sciences* (4 vols.), Thames and Hudson, London (from 1964).

(general editor), *Enseignement et diffusion des sciences en France au XVIIIe siècle*, Hermann, Paris (1964).

TEICH, Albert H., *International Politics and International Science: A Study of Scientists' Attitudes*, stencilled doctorate thesis, Department of Political Science, MIT, Cambridge, Mass. (1969).

TELLER, Edward, and BROWN, Allen, *The Legacy of Hiroshima*, Doubleday, New York (1962).

TOCQUEVILLE, Alexis de, *Democracy in America* (2 vols.) (trans. G. Lawrence), Fontana, London (1968).

TONDL, Ladislav, *Man and Science*, Institute for the Theory and Methodology of Science, Academy of Sciences, Prague (1969).

TYBOUT, R. A., (ed.), *Economics of Research and Development*, Ohio State University Press (1965).

VALLENTIN, Antonina, *Le drame d'Albert Einstein*, Plon, Paris (1954).

VAN DYKE, Vernon, *Pride and Power: The Rationale of the Space Program*, University of Illinois Press (1964).

WADDINGTON, C. H., The Scientific Attitude (1941: reissued by Hutchinson, London, 1968).

WATSON, James D., *The Double Helix*, Weidenfeld and Nicolson, London (1968; reissued by Penguin, Harmondsworth, 1970).

WEBER, Max, *Essays in Sociology* (ed. H. H. Gerth and C. W. Mills), Oxford University Press, New York (1970).

The Protestant Ethic and the Spirit of Captialism, Scribner, New York (1958).

The Methodology of the Social Sciences, Free Press of Glencoe (1964).

WEINBERG, Alvin M., *Reflections on Big Science*, MIT Press, Cambridge, Mass. (1967).

WHITEHEAD, A. N., *Science and the Modern World*, reissued by the New American Library, New York (1953).

WIENER, Norbert, *The Human Use of Human Beings—Cybernetics and Society*, Avon Books, New York (1967).

WIESNER, Jerome B., *Where Science and Politics Meet*, McGraw–Hill, New York (1965).

ZALESKI, E., KOZLOWSKI, J. P., WIENERT, H., DAVIES, R. W., BERRY, M. J., and AMANN, R., *Science Policy in the USSR*, OECD, Paris (1969).

Name Index

Subject Index